Collins

INTERNATIONAL PRIMARY MATHS

Progress Book 5

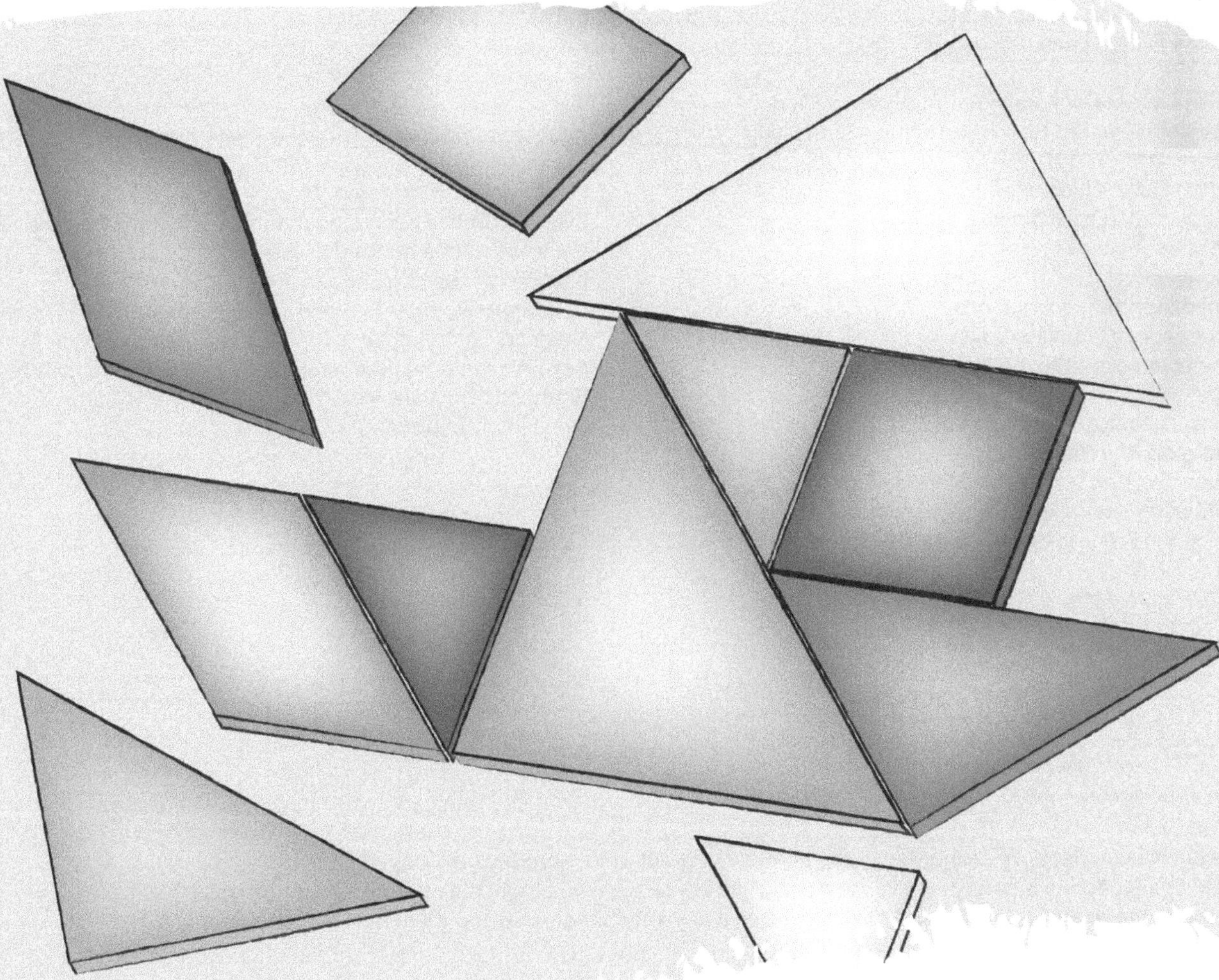

William Collins' dream of knowledge for all began with the publication of his first book in 1819. A self-educated mill worker, he not only enriched millions of lives, but also founded a flourishing publishing house. Today, staying true to this spirit, Collins books are packed with inspiration, innovation and practical expertise. They place you at the centre of a world of possibility and give you exactly what you need to explore it.

Collins. Freedom to teach.

Published by Collins
An imprint of HarperCollins*Publishers*
The News Building
1 London Bridge Street
London
SE1 9GF

HarperCollins*Publishers*
1st Floor Watermarque Building
Ringsend Road
Dublin 4
Ireland

Browse the complete Collins catalogue at
www.collins.co.uk

© HarperCollins*Publishers* Limited 2021

10 9 8 7 6 5 4 3 2

ISBN 978-0-00-836961-3

British Library Cataloguing-in-Publication Data
A catalogue record for this publication is available from the British Library.

Author: Peter Clarke
Series editor: Peter Clarke
Publisher: Elaine Higgleton
Product developer: Holly Woolnough
Copyeditor: Tanya Solomons
Proofreader: Catherine Dakin
Answer checker: Steven Matchett
Cover designer: Gordon MacGilp
Cover illustrator: Ann Paganuzzi
Illustrator: Ann Paganuzzi
Typesetter: Ken Vail Graphic Design Ltd
Production controller: Lyndsey Rogers
Printed and Bound in the UK using 100% Renewable Electricity at CPI Group (UK) Ltd

Photo acknowledgements
Every effort has been made to trace copyright holders. Any omission will be rectified at the first opportunity.
p15tl Volyk Nataliia/Shutterstock; p15tc Anatolir/Shutterstock; p15tr Liaksei Kruhlenia/Shutterstock; p15bl NeMaria/Shutterstock; p15bc Rafid/Shutterstock; p15br Asya Alexandrova/Shutterstock; p32 ANNA ZASIMOVA/Shutterstock; p44a Alexey Broslavets/Shutterstock; p44b Jemastock/Shutterstock.

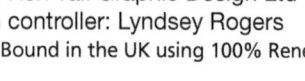 

MIX
Paper from
responsible sources
FSC™ C007454

This book is produced from independently certified FSC™ paper to ensure responsible forest management.

For more information visit: **www.harpercollins.co.uk/green**

The publishers gratefully acknowledge the permission granted to reproduce the copyright material in this book. Every effort has been made to trace copyright holders and to obtain their permission for the use of copyright material. The publishers will gladly receive any information enabling them to rectify any error or omission at the first opportunity.

Cambridge International copyright material in this publication is reproduced under licence and remains the intellectual property of Cambridge Assessment International Education

This text has not been through the Cambridge International endorsement process.

Contents

Introduction

The Progress Books include photocopiable end-of-unit progress tests which are designed to assist teachers with medium-term 'formative' assessment.

Each test is designed to be used within the classroom at the end of a Collins International Primary Maths unit to help measure the progress of learners and identify strengths and weaknesses.

Analysis of the results of the tests helps teachers provide feedback to individual learners on their specific strengths and areas that require improvement, as well as analyse the strengths and weaknesses of the class as a whole.

Self-assessment is also an important feature of the Progress Books as feedback should not only come from the teacher. The Progress Books provide opportunities at the end of each test for learners to self-assess their understanding of the unit, as well as space for teacher feedback.

Structure of the Progress Books

There is one progress test for each of the 27 units in Stage 5.

Each test consists of two pages of questions aimed at assessing the learning objectives from the Cambridge Primary Mathematics Curriculum Framework (0096) for the relevant unit. Where appropriate, this also includes questions to assess learners' development in one or more of the Thinking and Working Mathematically characteristics, as indicated by the TWM star on the page. All of the questions are typeset on triangles to indicate that they are suitable for the majority of learners and assess the unit's learning objectives.

Pages 5 to 9 include a list of photocopiable *I can* statements for each unit which are aimed at providing an opportunity for learners to undertake some form of self-assessment. The intention is that once learners have answered the two pages of questions, they turn to the *I can* statements for the relevant unit and think about each statement and how easy or hard they find the topic. For each statement they colour in the face that is closest to how they feel:

☺ I can do this   ☺ I'm getting there   ☹ I need some help.

A photocopiable variation of the Thinking and Working Mathematically Star is also included in the Progress Books. This version of the star includes *I can* statements for the eight TWM characteristics. Its purpose is to provide an opportunity for learners, twice a term/semester, to think about each of the statements and record how confident they feel about Thinking and Working Mathematically.

Administering each end-of-unit progress test

Recommended timing: 20 to 30 minutes, although this can be altered to suit the needs of individual learners and classes.

Before starting each end-of-unit progress test, ensure that each learner has the resources needed to complete the test. If needed, resources are listed in the 'You will need' box at the start of each test.

On completion of each end-of-unit progress test, use the answers and mark scheme available as a digital download to mark the tests.

Use the box at the bottom of the second page of each end-of-unit progress test to either:

- write the number of marks achieved by the learner out of the total marks possible.

- sign or initial your name to indicate you have marked the test.

- draw a simple picture or diagram such as one of the three faces (☺, ☺, ☹) to indicate your judgement on the learner's level of understanding of the unit's learning objectives.

- write a brief comment such as 'Well done!', 'You've got it', 'Getting there' or 'See me'.

Provide feedback to individual learners as necessary on their strengths and the areas that require improvement. Use the 'Class record-keeping document' located at the back of the Teacher's Guide and as a digital download to update your judgement of each learner's level of mastery in the relevant sub-strand.

I can statements

At the end of each unit, think about each of the *I can* statements and how easy or hard you find the topic. For each statement, colour in the face that is closest to how you feel.

Unit 1 – Counting and sequences	Date:			
• I can count on and back in steps of 7, 8 or 9.		☺	😐	☹
• I can find missing terms in a number sequence.		☺	😐	☹
• I understand square and triangular number sequences.		☺	😐	☹
Unit 2 – Addition of whole numbers	Date:			
• I can add combinations of 2- and 3-digit numbers.		☺	😐	☹
• I can add four 4-digit numbers, such as 7465 + 3626.		☺	😐	☹
• I can add a positive number to a negative number, such as $-5 + 8$.		☺	😐	☹
• I can solve missing number problems, such as $\square + 8 = 16$.		☺	😐	☹
• I can estimate the answer to an addition calculation.		☺	😐	☹
Unit 3 – Subtraction of whole numbers	Date:			
• I can subtract 4-digit numbers, such as 3837 – 2965.		☺	😐	☹
• I can subtract pairs of numbers where the answer is a negative number, such as $8 - 12$ and $-4 - 9$.		☺	😐	☹
• I can solve missing number problems, such as $\square - 7 = 15$.		☺	😐	☹
• I can estimate the answer to a subtraction calculation.		☺	😐	☹
Unit 4 – Multiples, factors, divisibility, primes and squares	Date:			
• I understand and can explain the difference between prime and composite numbers.		☺	😐	☹
• I can recognise numbers that are divisible by 4 and 8.		☺	😐	☹
• I can recognise square numbers from 1 to 100.		☺	😐	☹

Unit 5 – Whole number calculations	Date:		
• I know which property of number to use to simplify calculations.	☺	😐	☹
• I understand that the four operations follow a particular order.	☺	😐	☹
Unit 6 – Multiplication of whole numbers (A)	Date:		
• I can multiply numbers to 1000 by a 1-digit number.	☺	😐	☹
• I can estimate the answer to a multiplication calculation.	☺	😐	☹
Unit 7 – Multiplication of whole numbers (B)	Date:		
• I can multiply numbers to 1000 by a 2-digit number.	☺	😐	☹
• I can estimate the answer to a multiplication calculation.	☺	😐	☹
Unit 8 – Division of whole numbers (A)	Date:		
• I can divide numbers to 100 by a 1-digit number.	☺	😐	☹
• I can estimate the answer to a division calculation.	☺	😐	☹
Unit 9 – Division of whole numbers (B)	Date:		
• I can divide numbers to 1000 by a 1-digit number.	☺	😐	☹
• I can estimate the answer to a division calculation.	☺	😐	☹
Unit 10 – Place value and ordering decimals	Date:		
• I can explain the value of the tenths and hundredths digits in decimals.	☺	😐	☹
• I can compose and decompose decimals.	☺	😐	☹
• I can regroup decimals in different ways.	☺	😐	☹
• I can compare and order decimals.	☺	😐	☹
Unit 11 – Place value, ordering and rounding decimals	Date:		
• I can multiply and divide whole numbers by 10, 100 and 1000.	☺	😐	☹
• I can multiply decimals by 10 and 100.	☺	😐	☹
• I can divide decimals by 10.	☺	😐	☹
• I can round decimals to the nearest whole number.	☺	😐	☹

Unit 12 – Fractions (A)	Date:			
• I understand that a fraction can be represented by a division of the numerator by the denominator.		☺	😐	☹
• I can recognise improper fractions and mixed numbers.		☺	😐	☹
• I can convert between improper fractions and mixed numbers.		☺	😐	☹
• I can compare and order fractions with the same denominator.		☺	😐	☹
Unit 13 – Fractions (B)	**Date:**			
• I understand that proper fractions can act as operators.		☺	😐	☹
• I can add and subtract fractions with the same denominator.		☺	😐	☹
• I can add and subtract fractions with different denominators.		☺	😐	☹
• I can multiply unit fractions by a whole number.		☺	😐	☹
• I can divide unit fractions by a whole number.		☺	😐	☹
Unit 14 – Percentages	**Date:**			
• I can recognise percentages of shapes.		☺	😐	☹
• I can write percentages as a fraction with denominator 100.		☺	😐	☹
• I can compare and order percentages of quantities.		☺	😐	☹
Unit 15 – Addition and subtraction of decimals	**Date:**			
• I can add pairs of decimals mentally.		☺	😐	☹
• I can estimate and add pairs of decimals using a written method.		☺	😐	☹
• I can subtract pairs of decimals mentally.		☺	😐	☹
• I can estimate and subtract pairs of decimals using a written method.		☺	😐	☹
Unit 16 – Multiplication of decimals	**Date:**			
• I can multiply numbers with one decimal place by a 1-digit number mentally.		☺	😐	☹
• I can estimate and multiply numbers with one decimal place by a 1-digit number using a written method.		☺	😐	☹
Unit 17 – Fractions, decimals and percentages	**Date:**			
• I know that proper fractions, decimals and percentages can have equivalent values.		☺	😐	☹
• I can compare and order numbers with one decimal place, proper fractions and percentages.		☺	😐	☹

Unit 18 – Proportion and ratio	Date:			
• I understand that proportion compares part to whole.		☺	😐	☹
• I can describe proportions using fractions and percentages.		☺	😐	☹
• I understand that ratio compares part to part of two or more quantities.		☺	😐	☹
Unit 19 – Time	Date:			
• I understand time intervals of less than one second.		☺	😐	☹
• I can find time intervals in seconds, minutes and hours.		☺	😐	☹
• I can express time intervals as decimals, or in mixed units.		☺	😐	☹
• I can compare times between different time zones.		☺	😐	☹
Unit 20 – 2D shapes and symmetry	Date:			
• I can identify, describe, classify and sketch different triangles.		☺	😐	☹
• I can identify and create symmetrical patterns.		☺	😐	☹
Unit 21 – 3D shapes	Date:			
• I can identify, describe and sketch 3D shapes.		☺	😐	☹
• I can identify and sketch different nets for a cube.		☺	😐	☹
Unit 22 – Angles	Date:			
• I can estimate and classify angles.		☺	😐	☹
• I can compare angles.		☺	😐	☹
• I can calculate unknown angles on a straight line.		☺	😐	☹
Unit 23 – Perimeter and area	Date:			
• I can measure and calculate the perimeter of simple 2D shapes.		☺	😐	☹
• I can measure and calculate the perimeter of compound shapes.		☺	😐	☹
• I can measure and calculate the area of simple 2D shapes.		☺	😐	☹
• I can measure and calculate the area of compound shapes.		☺	😐	☹
• I understand that shapes with the same perimeter can have different areas and vice versa.		☺	😐	☹

Unit 24 – Coordinates	Date:			
• I can compare two points plotted on the coordinate grid to say which is closer to each axis.		☺	😐	☹
• Without a grid, I can estimate the position of point A relative to point B given the coordinates of point B.		☺	😐	☹
• I can plot points to form lines and shapes in the first quadrant.		☺	😐	☹
Unit 25 – Translation and reflection	Date:			
• I can describe and translate a shape on a square grid.		☺	😐	☹
• I can reflect 2D shapes in both horizontal and vertical mirror lines on square grids.		☺	😐	☹
• I can predict and draw where a shape will be after reflection where the sides of the shape are not vertical or horizontal.		☺	😐	☹
Unit 26 – Statistics	Date:			
• I can record, organise, represent and interpret data in a Venn or Carroll diagram to answer a question.		☺	😐	☹
• I can record, organise, represent and interpret data in a frequency table and bar chart to answer a question.		☺	😐	☹
• I can record, organise, represent and interpret data in a waffle diagram to answer a question.		☺	😐	☹
• I can find and interpret the mode and the median of a data set.		☺	😐	☹
Unit 27 – Statistics and probability	Date:			
• I can record, organise, represent and interpret data in a tally chart, frequency table, line graph or dot plot to answer a question.		☺	😐	☹
• I can describe the likelihood of an event happening.		☺	😐	☹
• I can describe the results of chance experiments.		☺	😐	☹

Number

Name: _____

1 Continue each sequence.

a 327, 334, 341, 348, 355, ☐ , ☐ , ☐ , ☐

b 244, 252, 260, 268, 276, ☐ , ☐ , ☐ , ☐

c 156, 147, 138, 129, 120, ☐ , ☐ , ☐ , ☐

d 337, 329, 321, 313, 305, ☐ , ☐ , ☐ , ☐

2 Write the missing numbers.

a 156, 165, ☐ , 183, 192, ☐ , 210, ☐ , ☐

b 483, 476, 469, ☐ , ☐ , 448, ☐ , 434, ☐

c 526, ☐ , 510, 502, ☐ , ☐ , 478, 470, ☐

d 285, 292, ☐ , ☐ , 313, 320, ☐ , 334, ☐

3 Continue each sequence.

a 7, 2, –3, –8, –13, ☐ , ☐ , ☐ , ☐

b –20, –12, –4, 4, 12, ☐ , ☐ , ☐ , ☐

c 39, 32, 25, 18, 11, ☐ , ☐ , ☐ , ☐

d –31, –27, –23, –19, –15, ☐ , ☐ , ☐ , ☐

4 Write the terms for each sequence.

a Count forward in 7s from 3.

3rd term ☐ 7th term ☐ 12th term ☐

b Count back in 8s from 12.

2nd term ☐ 6th term ☐ 10th term ☐

Number

 Circle the numbers that you would say when counting forward in 9s from 17.

24 37 44 52 62 71 81 90 98 109 115 125 139 143

 Write the missing numbers.

a   13, 10, ☐, 4, ☐, ☐, –5, ☐, –11

b   –25, –19, –13, ☐, ☐, ☐, 11, ☐, 23

c   –42, –33, ☐, –15, ☐, ☐, 12, 21, ☐

d   26, ☐, 12, ☐, –2, ☐, ☐, –23, –30

 Each of these sequences is linear, which means it increases or decreases by the same amount each time. Complete the missing terms.

a   43, ☐, ☐, ☐, ☐, 68

b   52, ☐, ☐, ☐, ☐, 32

c   182, ☐, ☐, ☐, ☐, 232

d   158, ☐, ☐, ☐, ☐, 143

 Write the next four numbers in each sequence.

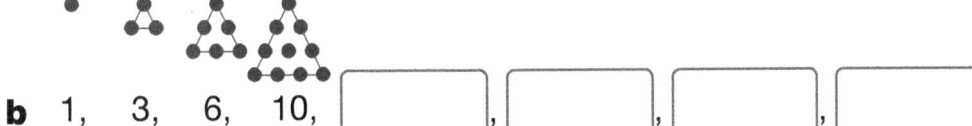

a   1,   4,   9,   16, ☐, ☐, ☐, ☐

b   1,   3,   6,   10, ☐, ☐, ☐, ☐

Now look at and think about each of the *I can* statements.

Date: _____

Number

Name:

1 Estimate, then use the most appropriate mental or written method to calculate the answer to each calculation. Show your working out.

a 46 + 85 + 319 =

Estimate:

b 235 + 326 + 127 =

Estimate:

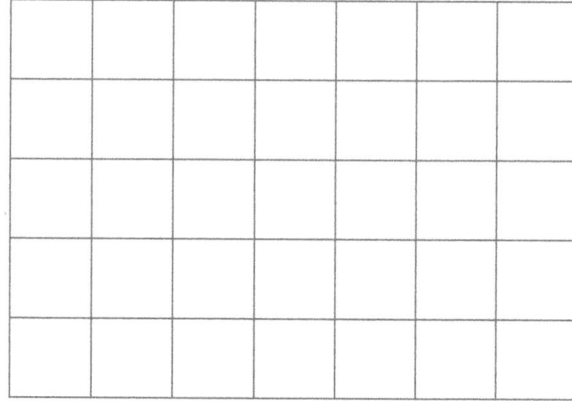

c 379 + 178 + 89 =

Estimate:

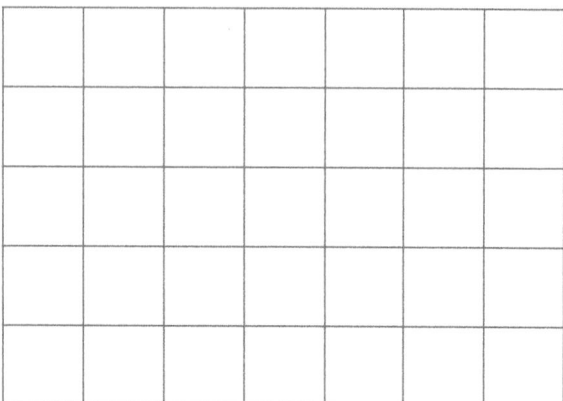

d 7364 + 8357 =

Estimate:

e 4633 + 2865 =

Estimate:

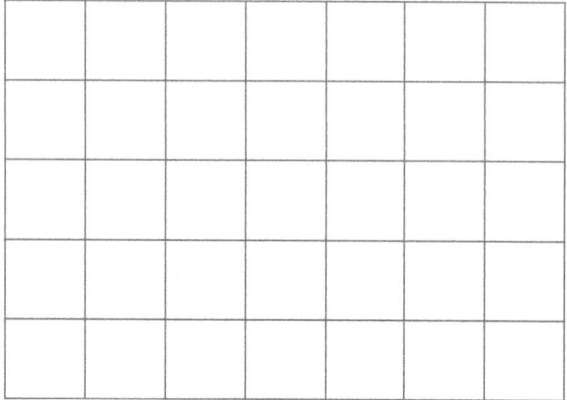

f 3276 + 5638 =

Estimate:

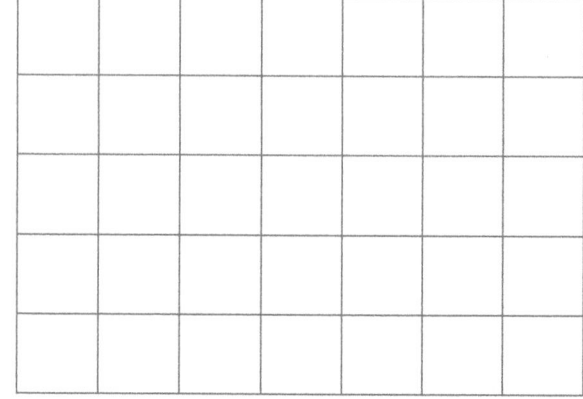

Number

2 Use the number line to find each answer.

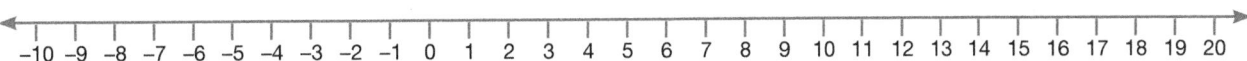

-10 -9 -8 -7 -6 -5 -4 -3 -2 -1 0 1 2 3 4 5 6 7 8 9 10 11 12 13 14 15 16 17 18 19 20

a $-4 + 6 = \boxed{}$

b $-5 + 10 = \boxed{}$

c $-2 + 9 = \boxed{}$

d $-9 + 4 = \boxed{}$

e $-8 + 7 = \boxed{}$

f $-3 + 11 = \boxed{}$

g $-1 + 8 = \boxed{}$

h $-10 + 5 = \boxed{}$

i $-6 + 12 = \boxed{}$

3 Use the number sentences to work out the price of each fruit.

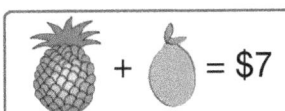

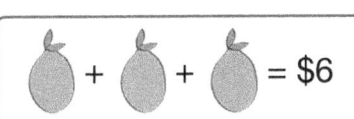

🍎 $= \boxed{}$ c

🍐 $= \$ \boxed{}$

🍇 $= \$ \boxed{}$

🍌 $= \$ \boxed{}$

🍍 $= \$ \boxed{}$

🍐 $= \$ \boxed{}$

4 Find the value of each symbol.

a $7 + \bigcirc = 13$

$\bigcirc = \boxed{}$

b $\square + 4 = 23$

$\square = \boxed{}$

c $\triangle + \triangle + \triangle = 15$

$\triangle = \boxed{}$

d $45 + \bigcirc = 60$

$\bigcirc = \boxed{}$

e $\square + \square = 80$

$\square = \boxed{}$

f $\triangle + 10 + \triangle = 50$

$\triangle = \boxed{}$

Now look at and think about each of the *I can* statements.

$\boxed{}$

Date: _____

13

Number

Name:
_____

 1 Estimate, then use the most appropriate mental or written method to calculate the answer to each calculation. Show your working out.

a 4726 – 2615 = [ ]

Estimate: [ ]

b 6483 – 2589 = [ ]

Estimate: [ ]

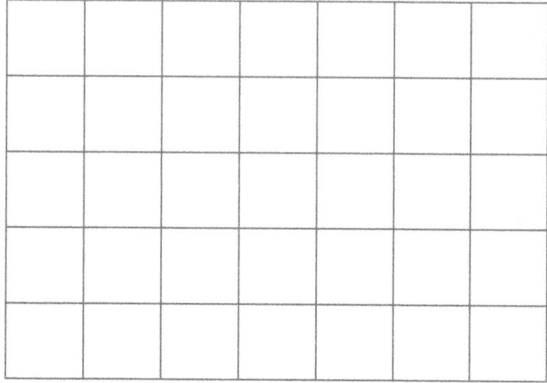

c 8372 – 4984 = [ ]

Estimate: [ ]

d 7362 – 4993 = [ ]

Estimate: [ ]

e 9583 – 9238 = [ ]

Estimate: [ ]

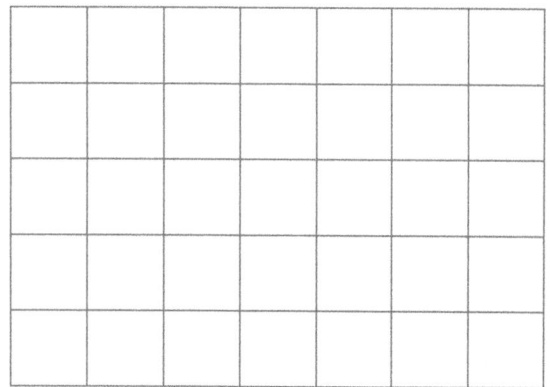

f 5033 – 1867 = [ ]

Estimate: [ ]

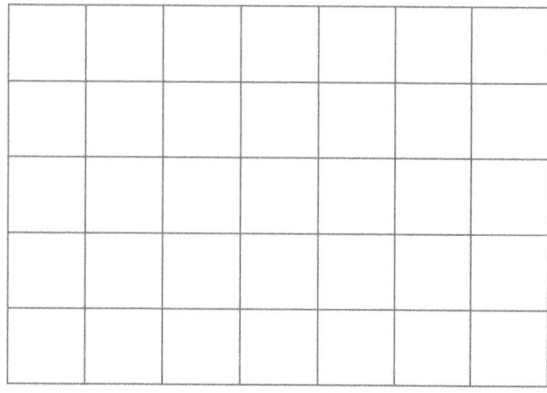

Number

2 Use the number line to find each answer.

```
←———————————————————————————————————————————→
 -10 -9 -8 -7 -6 -5 -4 -3 -2 -1  0  1  2  3  4  5  6  7  8  9 10 11 12 13 14 15 16 17 18 19 20
```

a $11 - 15 = \boxed{}$

b $6 - 9 = \boxed{}$

c $-2 - 5 = \boxed{}$

d $-3 - 6 = \boxed{}$

e $-5 - 5 = \boxed{}$

f $3 - 11 = \boxed{}$

g $1 - 8 = \boxed{}$

h $-6 - 1 = \boxed{}$

i $-4 - 2 = \boxed{}$

3 Use the number sentences to work out the price of each item.

$50 - = \$25$

$\$25 = \$35 - \text{⚽}$

$\text{🛹} - \$10 = \$10$

$\$20 = \text{🛼} - \text{⚽}$

$\$75 -  = \text{🖥}$

$\text{🤖} - \$5 = \$10$

⚽ $= \$ \boxed{}$

🛹 $= \$ \boxed{}$

🛴 $= \$ \boxed{}$

🖥 $= \$ \boxed{}$

🛼 $= \$ \boxed{}$

🤖 $= \$ \boxed{}$

4 Find the value of each symbol.

a $17 - \bigcirc = 6$

$\bigcirc = \boxed{}$

b $\square - 14 = 8$

$\square = \boxed{}$

c $16 = \triangle - 5$

$\triangle = \boxed{}$

d $100 - \bigcirc = 55$

$\bigcirc = \boxed{}$

e $\square - 30 = 70$

$\square = \boxed{}$

f $70 - \triangle - \triangle = 30$

$\triangle = \boxed{}$

Now look at and think about each of the *I can* statements.

$\boxed{}$

Date: _____

Number

Name: _____

You will need
• coloured pencils

 1 Write all the factors of each number.

a 12

b 16

c 20

 2 Draw a factor rainbow to find all the factors of each number.

a 72

b 56

3 For each number, circle to show if it is a prime number or a composite number. Then explain your answer.

a 19   prime number   composite number

_____

_____

b 39   prime number   composite number

_____

_____

4 Circle all the prime numbers.

9   11   17   21   23   26   29   30   31   35   37   41

42   43   53   57   61   66   67   71   79   89   92   97

5 Circle all the numbers that are divisible by 2.

147   352   427   603   838   945

1849   2671   5910   6125   8266   9364

6 How do you know a number is divisible by 2?

_____

_____

7 Circle all the numbers that are divisible by 4.

| 316 | 512 | 621 | 724 | 894 | 902 |

| 1818 | 2522 | 4442 | 5632 | 6920 | 7128 |

8 How do you know a number is divisible by 4?

_____

_____

9 Circle all the numbers that are divisible by 8.

| 128 | 364 | 482 | 536 | 842 | 952 |

| 2468 | 4825 | 4842 | 5376 | 7228 | 8736 |

10 How do you know a number is divisible by 8?

_____

_____

11 Write all the square numbers to 100.

Now look at and think about each of the *I can* statements.

Date: _____

Number

Name: _____

1 Solve each calculation mentally. Show any working out.

a $3 \times 6 \times 5 =$ ☐

b $9 \times 7 \times 2 =$ ☐

c $9 \times 5 \times 3 =$ ☐

d $2 \times 6 \times 4 =$ ☐

e $3 \times 3 \times 7 =$ ☐

f $8 \times 3 \times 2 =$ ☐

2 Solve each calculation mentally. Show any working out.

a $28 \times 7 =$ ☐

b $43 \times 9 =$ ☐

c $8 \times 64 =$ ☐

d $77 \times 6 =$ ☐

e $56 \times 9 =$ ☐

f $7 \times 89 =$ ☐

Number

3 Use the order of operations to calculate. Show any working out.

a $8 + 6 \times 5 =$ ☐

☐

b $18 \div 6 - 2 =$ ☐

☐

c $30 - 3 \times 8 =$ ☐

☐

d $6 \times 7 + 5 =$ ☐

☐

e $3 + 27 \div 3 =$ ☐

☐

f $14 - 24 \div 6 =$ ☐

☐

4 Write a calculation to represent each word problem and then solve it.

a A supermarket is selling cereal for $7 a packet. If you buy 3 packets you receive $2 off the total. What is the price of 3 packets of cereal?

☐

☐

b There are 6 cartons of milk in a box. A box of milk costs $18. Sam buys one carton of milk and a packet of cereal for $7. How much does Sam spend altogether?

☐

☐

Now look at and think about each of the *I can* statements.

☐

Date: _____

Number

Name: _____

1 Estimate, then use **partitioning** to calculate the answer to each calculation. Show your working out.

a 473 × 4 = [ ]

Estimate: [ ]

[ ] + [ ] + [ ]

= [ ] + [ ] + [ ]

= [ ]

b 629 × 8 = [ ]

Estimate: [ ]

[ ] + [ ] + [ ]

= [ ] + [ ] + [ ]

= [ ]

2 Estimate, then use the **grid method** to calculate the answer to each calculation. Show your working out.

a 584 × 7 = [ ]

Estimate: [ ]

× [ ] [ ] [ ]

[ ] [ ] [ ] [ ]

[ ]

b 736 × 9 = [ ]

Estimate: [ ]

× [ ] [ ] [ ]

[ ] [ ] [ ] [ ]

[ ]

3 Estimate, then use the **expanded written method** to calculate the answer to each calculation. Show your working out.

a Estimate: [ ]

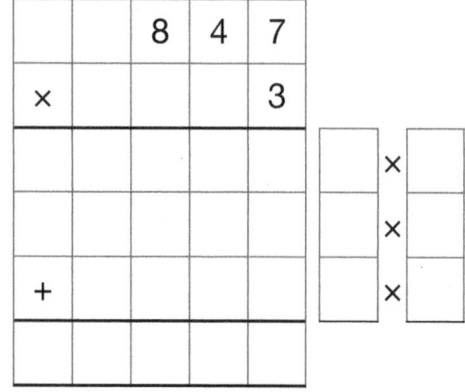

b Estimate: [ ]

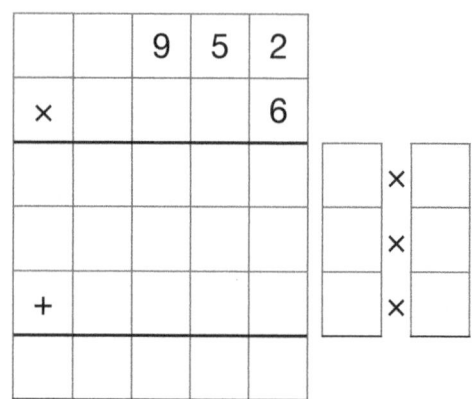

4 Estimate, then use your **preferred method** to work out the answer to each calculation. Show your working out.

a 368 × 7 = [   ]

Estimate: [   ]

b 894 × 3 = [   ]

Estimate: [   ]

5 Work out the answer to each problem. Show your working out.

a Mika works 6 days a week picking apples. He picks around 825 apples each day. How many apples does Mika pick in 6 days?

[   ]

b In one day, 648 bags of pears are packed. Each bag holds 8 pears. How many pears were packed in the day?

[   ]

Now look at and think about each of the *I can* statements.

[   ]

Date: _____

Number

Name: _____

 Estimate, then use the **area diagram** to calculate the answer to each calculation. Show your working out.

a $24 \times 18 =$ [ ]

Estimate: [ ]

× [ ][ ]
[ ][ ]

[ ]

b $36 \times 27 =$ [ ]

Estimate: [ ]

× [ ][ ]
[ ][ ]

[ ]

 Estimate, then use the **grid method** to calculate the answer to each calculation. Show your working out.

a $53 \times 42 =$ [ ]

Estimate: [ ]

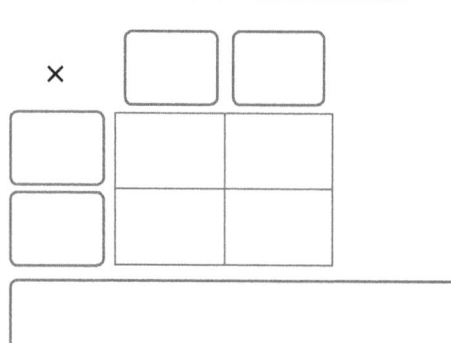

[ ]

b $68 \times 39 =$ [ ]

Estimate: [ ]

[ ]

3 Work out the answer to the problem. Show your working out.

An orchestra sells 84 tickets to a concert. Each ticket costs $26. How much money does the orchestra make?

Number

4 Estimate, then use the **expanded written method** to calculate the answer to each calculation. Show your working out.

a Estimate: [   ]

			6	3
×			5	8
+				

× [ ]
× [ ]
× [ ]
× [ ]

b Estimate: [   ]

		3	5	2
×			2	4
+				

× [ ]
× [ ]
× [ ]
× [ ]
× [ ]
× [ ]

5 Estimate, then use your **preferred method** to work out the answer to each calculation. Show your working out.

a 45 × 86 = [   ]

Estimate: [   ]

[   ]

b 463 × 32 = [   ]

Estimate: [   ]

[   ]

Now look at and think about each of the *I can* statements.

[   ]

Date: _____

Number

Name: _____

 Estimate, then use **partitioning** to calculate the answer to each calculation. Show your working out.

a 96 ÷ 4 = ☐

Estimate: ☐

96 ÷ 4 = ☐ + ☐ ÷ ☐

= ☐ ÷ ☐ + ☐ ÷ ☐

= ☐ + ☐

= ☐

b 81 ÷ 3 = ☐

Estimate: ☐

81 ÷ 3 = ☐ + ☐ ÷ ☐

= ☐ ÷ ☐ + ☐ ÷ ☐

= ☐ + ☐

= ☐

2 Estimate, then use **partitioning** to calculate the answer to each calculation. Show your working out.

a 91 ÷ 7 = ☐

Estimate: ☐

b 96 ÷ 6 = ☐

Estimate: ☐

3 Work out the answer to the problem. Show your working out.

76 children are arranged into teams of 4. How many teams are there?

☐

Number

4 Estimate, then use the **expanded written method** to calculate the answer to each calculation. Show your working out.

a 84 ÷ 3 = [ ]

Estimate: [ ]

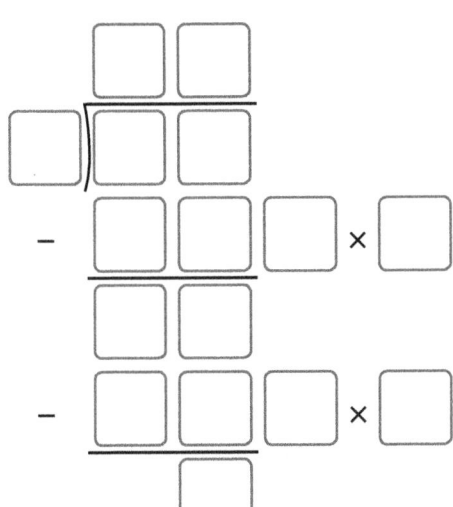

b 82 ÷ 6 = [ ]

Estimate: [ ]

5 Estimate, then use your **preferred method** to work out the answer to each calculation. Show your working out.

a 72 ÷ 4 = [ ]

Estimate: [ ]

b 97 ÷ 5 = [ ]

Estimate: [ ]

Now look at and think about each of the *I can* statements. [ ]

Date: _____

Number

Name: _____

1 Estimate, then use **partitioning** to calculate the answer to each calculation. Show your working out.

a 288 ÷ 8 = [ ]

Estimate: [ ]

288 ÷ 8 = [+] ÷ [ ]

= [÷] + [÷]

= [+]

= [ ]

b 204 ÷ 3 = [ ]

Estimate: [ ]

204 ÷ 3 = [+] ÷ [ ]

= [÷] + [÷]

= [+]

= [ ]

2 Estimate, then use **partitioning** to calculate the answer to each calculation. Show your working out.

a 360 ÷ 5 = [ ]

Estimate: [ ]

b 486 ÷ 6 = [ ]

Estimate: [ ]

3 Work out the answer to the problem. Show your working out.

On one day, a canoe company carries a total of 522 passengers. On each trip, it carries 9 passengers. How many trips does it make?

[ ]

4 Estimate, then use the **expanded written method** to calculate the answer to each calculation. Show your working out.

a $376 \div 4 =$ ☐

Estimate: ☐

b $611 \div 8 =$ ☐

Estimate: ☐

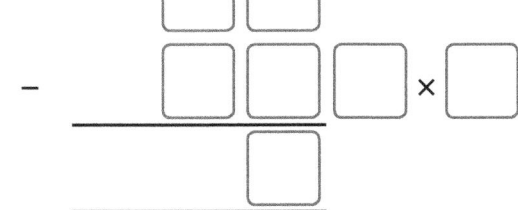

5 Estimate, then use your **preferred method** to work out the answer to each calculation. Show your working out.

a $546 \div 7 =$ ☐

Estimate: ☐

b $786 \div 9 =$ ☐

Estimate: ☐

Now look at and think about each of the *I can* statements. ☐

Date: _____

© HarperCollins*Publishers* Limited 2021

27

Number

Name: _____

1 Write the decimal that is equivalent to each fraction.

a $\frac{4}{10}$ = ☐

b $\frac{9}{10}$ = ☐

c $\frac{18}{100}$ = ☐

d $\frac{37}{100}$ = ☐

e $\frac{6}{100}$ = ☐

f $\frac{51}{100}$ = ☐

2 Decompose each number by place value.

a 0·48 = ☐ + ☐

b 0·23 = ☐ + ☐

c 5·16 = ☐ + ☐ + ☐

d 9·57 = ☐ + ☐ + ☐

3 Complete each statement.

a 67·35 is composed from ☐ ☐ ☐ ☐

b 321·94 is composed from ☐ ☐ ☐ ☐ ☐

c 48·12 is composed from ☐ ☐ ☐ ☐

d 763·94 is composed from ☐ ☐ ☐ ☐ ☐

4 Write each decimal.

a 4 ones, 7 tenths and 2 hundredths
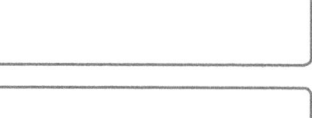

b 8 tens, 3 ones, 7 tenths and 5 hundredths

c 2 hundredths, 4 tens, 9 ones and 3 tenths

d 7 tenths, 1 ten, 4 hundredths and 8 ones
☐

Number

5 Decompose each number in four different ways.

a
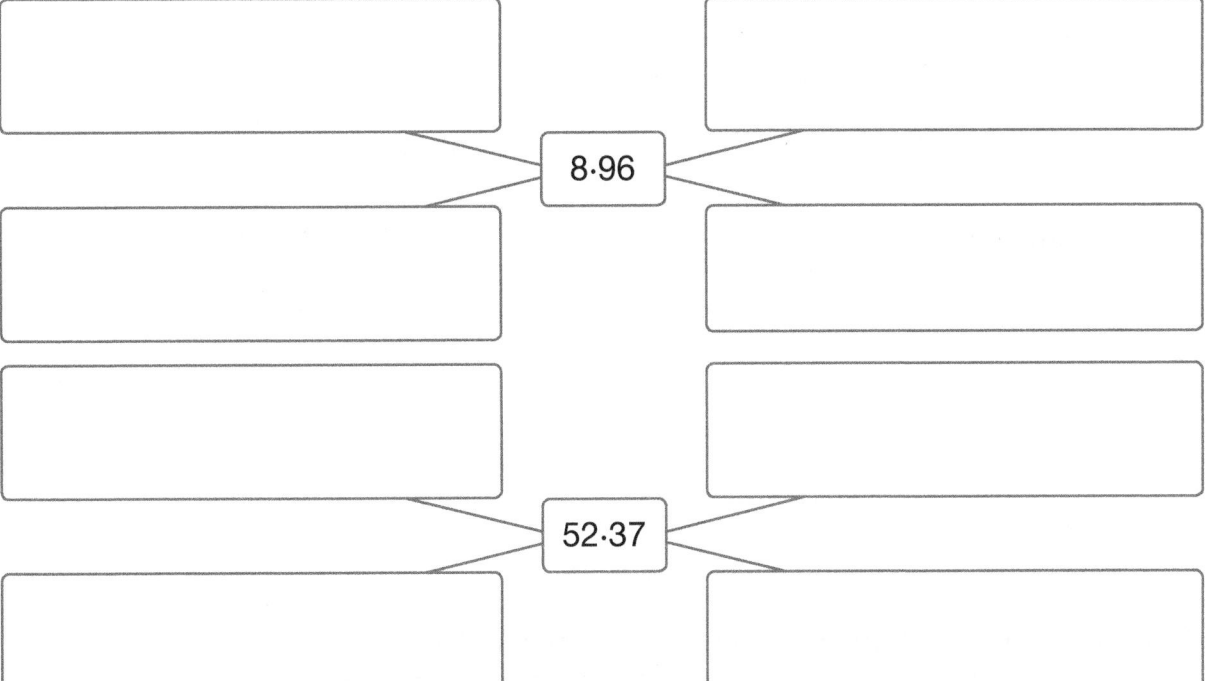

b

6 Write the correct symbol, < or >, to compare each pair of decimals.

a 0·5 [ ] 0·7

b 4·8 [ ] 4·1

c 12·3 [ ] 12·9

d 43·2 [ ] 42·3

e 13·3 [ ] 13·2

f 42·3 [ ] 42·4

7 Order each set of decimals, **smallest** first.

a 0·5, 0·3, 0·8, 0·6, 0·2 [ ], [ ], [ ], [ ], [ ]

b 5·3, 4·2, 6·1, 3·8, 2·9 [ ], [ ], [ ], [ ], [ ]

c 24·3, 43·2, 34·2, 23·4, 42·3 [ ], [ ], [ ], [ ], [ ]

d 67·3, 67·5, 67·2, 67·8, 67·7 [ ], [ ], [ ], [ ], [ ]

Now look at and think about each of the *I can* statements. [ ]

Date: _____

Number

Name: _____

1 Complete each calculation.

a 58 × 10 = [ ]

b 82 ÷ 100 = [ ]

c 420 ÷ 10 = [ ]

d 3 × 1000 = [ ]

e 387 × 10 = [ ]

f 470 ÷ 1000 = [ ]

g 6 × 100 = [ ]

h 4 ÷ 10 = [ ]

i 5323 ÷ 100 = [ ]

j 81 × 100 = [ ]

k 254 × 100 = [ ]

l 30 ÷ 1000 = [ ]

2 Complete the missing numbers in the place value charts.

a

1000s	100s	10s	1s	.	$\frac{1}{10}$s	$\frac{1}{100}$s	
		6	8	.	2		
				.			× 10
				.			× 100

b

1000s	100s	10s	1s	.	$\frac{1}{10}$s	$\frac{1}{100}$s	
		5	1	.	3	9	
				.			× 10
				.			× 100

3 Complete each calculation.

a 14·3 × 10 = [ ]

b 1·6 × 100 = [ ]

c 25·9 × 100 = [ ]

d 8·34 × 10 = [ ]

e 12·97 × 10 = [ ]

f 32·75 × 100 = [ ]

g 9·34 × 100 = [ ]

h 435·5 × 10 = [ ]

Number

 4 Complete the missing numbers in the place value charts.

a

1000s	100s	10s	1s	$\frac{1}{10}$s	$\frac{1}{100}$s	
			5	2		
						÷ 10

b

1000s	100s	10s	1s	$\frac{1}{10}$s	$\frac{1}{100}$s	
	2	3	4	1		
						÷ 10

 5 Complete each calculation.

a 70·8 ÷ 10 = ☐

b 6·2 ÷ 10 = ☐

c 451·9 ÷ 10 = ☐

d 0·7 ÷ 10 = ☐

e 15·4 ÷ 10 = ☐

f 205·8 ÷ 10 = ☐

g 200·5 ÷ 10 = ☐

h 42·6 ÷ 10 = ☐

 6 Write the whole number on either side of each decimal.
Then circle the whole number that the decimal rounds to.

a ☐ 5·8 ☐

b ☐ 16·2 ☐

c ☐ 43·6 ☐

d ☐ 9·4 ☐

7 Round each measurement to the nearest whole unit.

a 5·7 kg ☐ kg

b 7·3 *l* ☐ *l*

c 13·8 m ☐ m

d 26·1 km ☐ km

e 57·9 g ☐ g

f 9·5 cm ☐ cm

Now look at and think about each of the *I can* statements. ☐

Date: _____

Number

Name: _____

1 Express each fraction as a division calculation.

a $\frac{1}{3}$ [ ] ÷ [ ]

b $\frac{7}{10}$ [ ] ÷ [ ]

c $\frac{1}{100}$ [ ] ÷ [ ]

d $\frac{3}{4}$ [ ] ÷ [ ]

e $\frac{1}{10}$ [ ] ÷ [ ]

f $\frac{47}{100}$ [ ] ÷ [ ]

2 Colour each mixed number.

a $2\frac{3}{4}$

b $1\frac{2}{5}$

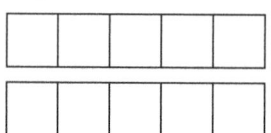

3 Express each diagram as a mixed number.

a

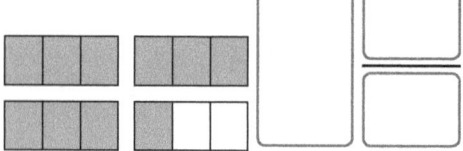

b

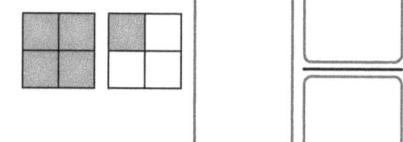

4 Colour each improper fraction, then write the fraction as a mixed number.

a $\frac{11}{3}$ =

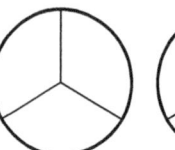 

b $\frac{23}{5}$ =

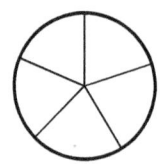 

c $\frac{27}{8}$ =

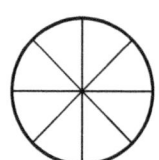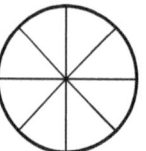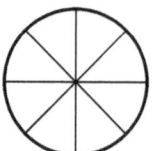 

d $\frac{18}{4}$ =

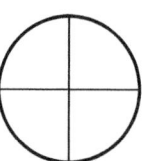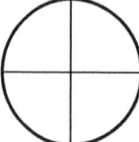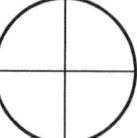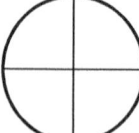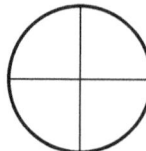 

5 Convert each mixed number to an improper fraction.
Show your working out.

a $2\frac{3}{5} =$ ▢

b $1\frac{2}{3} =$ ▢

c $3\frac{1}{4} =$ ▢

d $4\frac{1}{2} =$ ▢

e $6\frac{7}{10} =$ ▢

f $5\frac{5}{6} =$ ▢

6 Write the correct symbol, < or >, to compare each pair of fractions.

a $\frac{3}{5}$ ▢ $\frac{1}{5}$

b $\frac{1}{4}$ ▢ $\frac{3}{4}$

c $\frac{7}{10}$ ▢ $\frac{4}{10}$

d $\frac{5}{7}$ ▢ $\frac{3}{7}$

e $\frac{4}{9}$ ▢ $\frac{5}{9}$

f $\frac{2}{6}$ ▢ $\frac{5}{6}$

7 Order each set of fractions, **smallest** first.

a $\frac{4}{8}$ $\frac{2}{8}$ $\frac{1}{8}$ $\frac{7}{8}$ $\frac{6}{8}$ $\frac{3}{8}$

▢ , ▢ , ▢ , ▢ , ▢ , ▢

b $\frac{8}{10}$ $\frac{5}{10}$ $\frac{2}{10}$ $\frac{4}{10}$ $\frac{10}{10}$ $\frac{6}{10}$

▢ , ▢ , ▢ , ▢ , ▢ , ▢

Now look at and think
about each of the
I can statements.

Date: _____

Number

Name: _____

1 Work out these fractions. Show your working out.

a $\frac{3}{10}$ of 60 = ☐

b $\frac{7}{10}$ of 40 = ☐

c $\frac{4}{100}$ of 800 = ☐

d $\frac{9}{100}$ of 200 = ☐

2 Solve these fraction calculations.

a $\frac{4}{7} + \frac{2}{7} = \frac{\boxed{}}{\boxed{}}$

b $\frac{4}{5} - \frac{2}{5} = \frac{\boxed{}}{\boxed{}}$

c $\frac{3}{8} + \frac{2}{8} = \frac{\boxed{}}{\boxed{}}$

d $\frac{7}{9} - \frac{3}{9} = \frac{\boxed{}}{\boxed{}}$

e $\frac{4}{10} + \frac{6}{10} = \frac{\boxed{}}{\boxed{}}$

f $\frac{11}{12} - \frac{5}{12} = \frac{\boxed{}}{\boxed{}}$

3 Use the fraction wall to solve these fraction calculations.

a $\frac{1}{2} + \frac{1}{4} = \frac{\boxed{}}{\boxed{}}$

b $\frac{3}{4} - \frac{1}{2} = \frac{\boxed{}}{\boxed{}}$

c $\frac{5}{6} - \frac{1}{3} = \frac{\boxed{}}{\boxed{}}$

d $\frac{2}{3} + \frac{1}{6} = \frac{\boxed{}}{\boxed{}}$

e $\frac{7}{8} - \frac{3}{4} = \frac{\boxed{}}{\boxed{}}$

f $\frac{5}{6} + \frac{2}{3} = \frac{\boxed{}}{\boxed{}}$

1							
$\frac{1}{2}$				$\frac{1}{2}$			
$\frac{1}{3}$		$\frac{1}{3}$			$\frac{1}{3}$		
$\frac{1}{4}$		$\frac{1}{4}$		$\frac{1}{4}$		$\frac{1}{4}$	
$\frac{1}{6}$	$\frac{1}{6}$	$\frac{1}{6}$	$\frac{1}{6}$	$\frac{1}{6}$	$\frac{1}{6}$		
$\frac{1}{8}$	$\frac{1}{8}$	$\frac{1}{8}$	$\frac{1}{8}$	$\frac{1}{8}$	$\frac{1}{8}$	$\frac{1}{8}$	$\frac{1}{8}$

4 Use the diagrams to multiply.

a $\frac{1}{2} \times 3 = \frac{\square}{\square}$

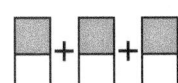

b $\frac{1}{4} \times 3 = \frac{\square}{\square}$

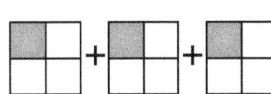

c $\frac{1}{5} \times 4 = \frac{\square}{\square}$

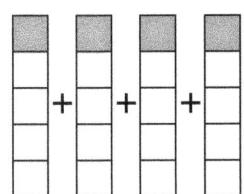

d $\frac{1}{6} \times 5 = \frac{\square}{\square}$

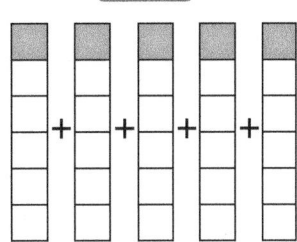

 5 Divide these fractions. Draw models to help you.

a $\frac{1}{2} \div 2 = \frac{\square}{\square}$

b $\frac{1}{5} \div 4 = \frac{\square}{\square}$

c $\frac{1}{4} \div 3 = \frac{\square}{\square}$

d $\frac{1}{3} \div 5 = \frac{\square}{\square}$

Now look at and think about each of the *I can* statements.

Date: _____

Number

Name:
_____

You will need
• coloured pencil

 1 The shaded part of each 100 grid represents a percentage. Write the percentage shown.

a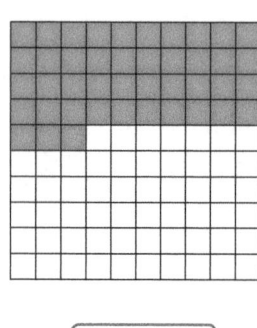

b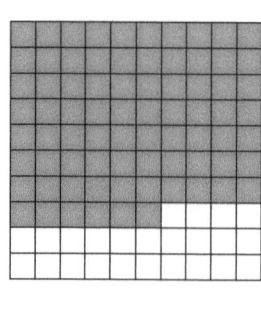

c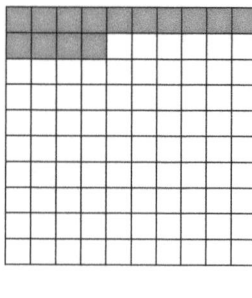

[____] %   [____] %   [____] %

 2 Shade each grid to show the percentage given.

a 55%   b 82%   c 27%

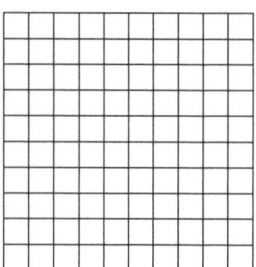   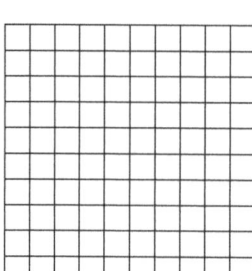   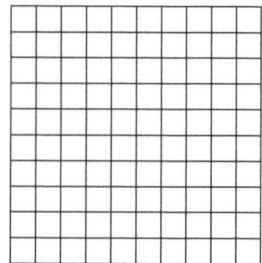

3 Shade the fraction of the grid shown and write the percentage.

a $\frac{1}{10}$ = [____] %   b $\frac{1}{4}$ = [____] %   c $\frac{16}{100}$ = [____] %

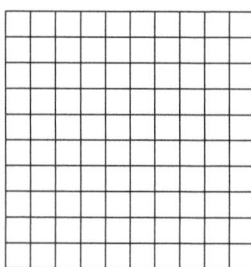   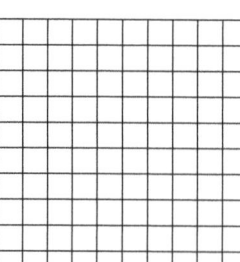   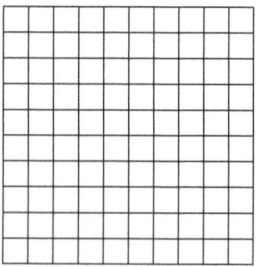

 4 Write each statement as a percentage and a fraction.

a 39 out of every 100 apples are green.   b 91 out of every 100 pears are ripe.

[____] % = [__]/[__]   [____] % = [__]/[__]

36

5 Write the equivalent fraction for each percentage.

a 62% = $\dfrac{\boxed{}}{\boxed{}}$

b 19% = $\dfrac{\boxed{}}{\boxed{}}$

c 48% = $\dfrac{\boxed{}}{\boxed{}}$

6 Write the equivalent percentage for each fraction.

a $\frac{3}{4}$ = $\boxed{}$%

b $\frac{9}{10}$ = $\boxed{}$%

c $\frac{31}{100}$ = $\boxed{}$%

7 Mark each pair of numbers on the number line. Then complete the statement with < or >.

a 30% $\boxed{}$ $\frac{4}{10}$

b $\frac{9}{10}$ $\boxed{}$ 80%

c 70% $\boxed{}$ $\frac{5}{10}$

8 Mark the numbers on the number line. Then write the numbers in order, from smallest to greatest.

a $\frac{7}{10}$ 50% $\frac{2}{10}$

$\boxed{}$ < $\boxed{}$ < $\boxed{}$

b 60% $\frac{3}{4}$ 40%

$\boxed{}$ < $\boxed{}$ < $\boxed{}$

c 30% $\frac{1}{10}$ 70% $\frac{1}{2}$

$\boxed{}$ < $\boxed{}$ < $\boxed{}$ < $\boxed{}$

Now look at and think about each of the *I can* statements.

Date: _____

© HarperCollins*Publishers* Limited 2021

37

Number

Name: _____

1 Solve these calculations **mentally** using your preferred strategies.
Show your working out.

a 2·6 + 4·8 = [ ]

b 14·6 + 18·7 = [ ]

c 12·5 + 18·6 = [ ]

d 1·34 + 4·67 = [ ]

e 8·54 + 6·31 = [ ]

f 9·57 + 2·84 = [ ]

2 Estimate, then solve each calculation using the **expanded written method**. Show your working out.

a 34·67 + 16·74 = [ ]

Estimate: [ ]

b 52·83 + 27·92 = [ ]

Estimate: [ ]

Number

3 Estimate, then solve each calculation using the **formal written method**. Show your working out.

a 43·76 + 17·58 = ⬚

Estimate: ⬚

			.		
+			.		
			.		

b 48·32 + 31·88 = ⬚

Estimate: ⬚

			.		
+			.		
			.		

4 Solve these calculations **mentally** using your preferred strategies. Show your working out.

a 3·8 − 2·6 = ⬚

b 9·4 − 2·5 = ⬚

c 73·5 − 9·4 = ⬚

d 52·5 − 14·7 = ⬚

5 Estimate, then solve each calculation using the **formal written method**. Show your working out.

a 74·38 − 35·64 = ⬚

Estimate: ⬚

			.		
−			.		
			.		

b 51·53 − 26·78 = ⬚

Estimate: ⬚

			.		
−			.		
			.		

Now look at and think about each of the *I can* statements.

⬚

Date: _____

39

Number

Name: _____

 Use your knowledge of times table facts to find the products.

a 0·6 × 4 = [   ]

b 0·3 × 7 = [   ]

c 0·8 × 6 = [   ]

d 0·5 × 9 = [   ]

e 0·4 × 8 = [   ]

f 0·7 × 5 = [   ]

 Use **partitioning** to work out the answer to each calculation.
Show your working out.

a 4·6 × 7 = [   ]

b 5·3 × 6 = [   ]

c 8·7 × 9 = [   ]

d 9·7 × 8 = [   ]

 Estimate, then use the **grid method** to work out the answer to each calculation.
Show your working out.

a 23·6 × 7 = [   ]

Estimate: [   ]

× [ ] [ ] [ ]

[ ] [ ] [ ]

[   ]

b 38·4 × 9 = [   ]

Estimate: [   ]

× [ ] [ ] [ ]

[ ] [ ] [ ]

[   ]

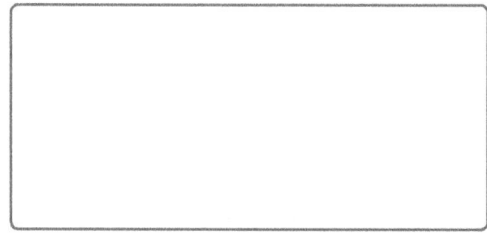

Number

4 Estimate, then use the **expanded written method** to work out the answer to each calculation. Show your working out.

a 54·7 × 6 = [   ] × 6 ÷ 10

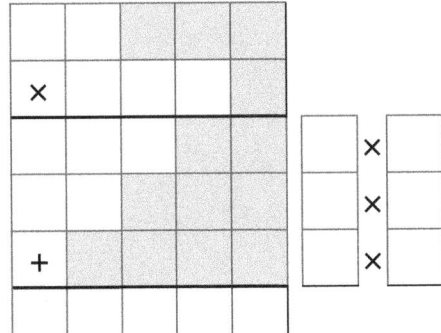

Answer: [   ] ÷ 10 = [   ]

b 47·3 × 8 = [   ] × 8 ÷ 10

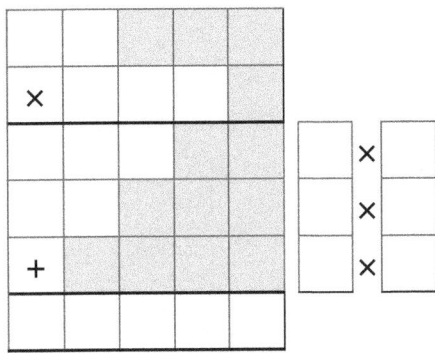

Answer: [   ] ÷ 10 = [   ]

5 Estimate, then use your **preferred method** to work out the answer to each calculation. Show your working out.

a 5·6 × 7 = [   ]

Estimate: [   ]

b 8·2 × 6 = [   ]

Estimate: [   ]

c 42·3 × 8 = [   ]

Estimate: [   ]

d 71·4 × 9 = [   ]

Estimate: [   ]

Now look at and think about each of the *I can* statements.

Date: _____

Number

Name: _____

You will need
• coloured pencil

1 Shade each number on the 100 grid and write the missing fraction, percentage and decimal equivalents.

a $\frac{4}{10}$ = [ ] % = [·]

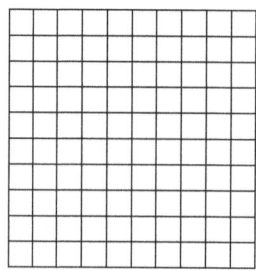

b $\frac{[\ ]}{[\ ]}$ = [ ] % = 0·2

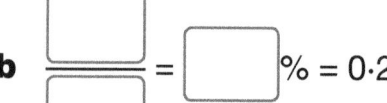

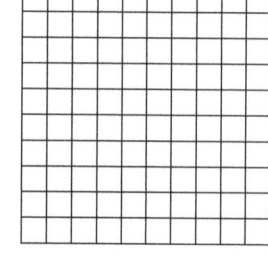

c $\frac{70}{100}$ = [ ] % = [·]

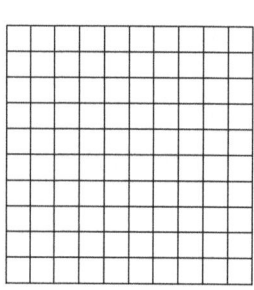

d $\frac{[\ ]}{[\ ]}$ = 25% = [·]

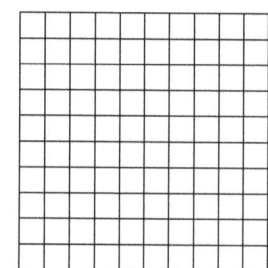

2 Complete the equivalent representations for each number.

a $\frac{30}{100}$ = [ ] % = [·]

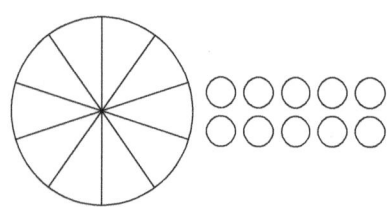

b $\frac{[\ ]}{[\ ]}$ = 10% = [·]

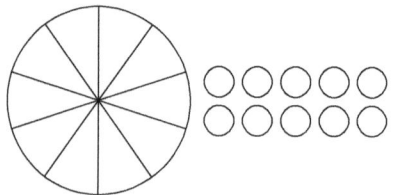

c $\frac{9}{10}$ = [ ] % = [·]

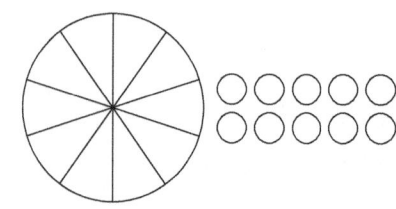

d $\frac{[\ ]}{[\ ]}$ = [ ] % = 0·5

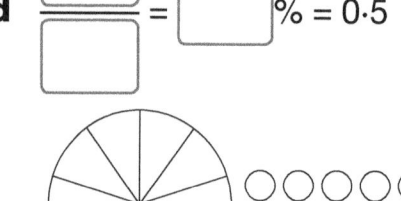

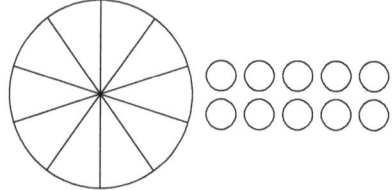

Number

 3 Use <, > and = to compare the numbers.

a 30% ☐ 0·4   **b** $\frac{3}{5}$ ☐ 80%   **c** 0·25 ☐ $\frac{3}{4}$

d $\frac{2}{10}$ ☐ 0·1   **e** 0·5 ☐ 50%   **f** 90% ☐ $\frac{8}{10}$

g 70% ☐ 0·9   **h** 0·6 ☐ $\frac{2}{5}$ **i** $\frac{1}{4}$ ☐ 25%

 4 Circle the **largest** number in each row.

a $\frac{1}{5}$ 10%   0·25

b 0·75 $\frac{1}{4}$ 50%

c 20%   0·4 $\frac{30}{100}$

 5 Order each set of numbers, starting with the **smallest**.

a   0·3   40% $\frac{2}{10}$ 0·6   50%

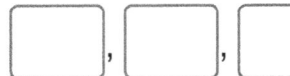
☐ , ☐ , ☐ , ☐ , ☐

b $\frac{3}{10}$ 0·2   10%   0·25 $\frac{1}{2}$

☐ , ☐ , ☐ , ☐ , ☐

c   25%   0·1 $\frac{1}{5}$ 30% $\frac{4}{10}$

☐ , ☐ , ☐ , ☐ , ☐

 6 Position the numbers on the number line. Circle the **smallest** number.

a 60% $\frac{1}{2}$ 0·3

0 ——————————————— 1

b 0·8 40% $\frac{7}{10}$

0 ——————————————— 1

Now look at and think about each of the *I can* statements.

☐

Date: _____

Number

Name: _____

1 Complete the sentences.

a

[ ] in every [ ] balloons have stripes.

The fraction of balloons that have stripes is $\dfrac{[\quad]}{[\quad]}$.

b

[ ] in every [ ] balloons have stripes.

The fraction of balloons that have stripes is $\dfrac{[\quad]}{[\quad]}$.

2 In every 10 apples in a crate, there are 4 green apples and 6 red apples. Write the proportion of green and red apples in a crate as a fraction and as a percentage.

Green apples: $\dfrac{[\quad]}{[\quad]}$ [ ] % Red apples: $\dfrac{[\quad]}{[\quad]}$ [ ] %

3 In every 20 street lights, 15 are working. The rest are broken. Write the proportion of lights that are working and those that are not as a fraction and as a percentage.

Working: $\dfrac{[\quad]}{[\quad]}$ [ ] % Broken: $\dfrac{[\quad]}{[\quad]}$ [ ] %

4 Write the ratios. Simplify them where possible.

a In a class of 18 children, 8 are boys and 10 are girls.

Boys to girls: [ ] : [ ]

b In a sports team, there are 5 girls for every 3 boys.

Girls to boys: [ ] : [ ]

5 Write the ratios. Simplify them where possible.

a

Rabbits to carrots: ☐ : ☐   Carrots to rabbits: ☐ : ☐

b

Rockets to stars: ☐ : ☐   Stars to rockets: ☐ : ☐

c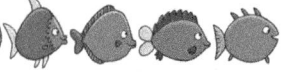

Fish to shells: ☐ : ☐   Shells to fish: ☐ : ☐

6 Write the ratio and proportion of **circles to triangles**.

a ▲ ▲ ▲ ● ● ● ● ● ●   Ratio ☐ : ☐

Proportion of circles

☐/☐   ☐ %

Proportion of triangles

☐/☐   ☐ %

b ▲ ▲ ▲ ▲ ● ● ● ● ● ●   Ratio ☐ : ☐

Proportion of circles

☐/☐   ☐ %

Proportion of triangles

☐/☐   ☐ %

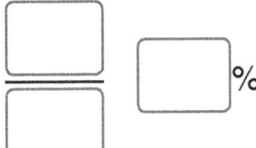

Now look at and think about each of the *I can* statements.

☐

Date: _____

Geometry and Measure

Name: _____

1 Write something that takes about a second or less.

_____

_____

2 Calculate the amount of time that has lapsed between each pair of clocks. Show your working out.

a

○ a.m.   ● p.m.   ○ a.m.   ● p.m.

b

● a.m.   ○ p.m.   ● a.m.   ○ p.m.

c

● a.m.   ○ p.m.   ○ a.m.   ● p.m.

d
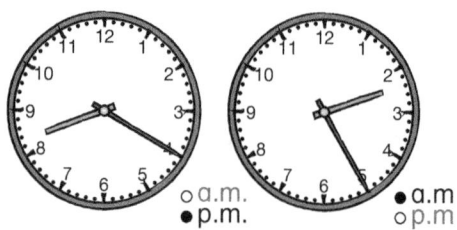
○ a.m.   ● p.m.   ● a.m.   ○ p.m

3 Convert these decimal times.

a 0·5 h = ☐ min

b 0·25 min = ☐ s

c 2·5 days = ☐ h

d 0·5 min = ☐ s

e 0·75 h = ☐ min

f 1·25 days = ☐ h

Geometry and Measure

4 Four children practise the piano. Calculate the amount of time each child spent. Show any working out in the last column.

Child	Start time	End time	Time spent (min)
Gopal	16:36	17:04	
Sunita	16:20	17:18	
Lakshmi	18:53	19:11	
Krish	17:42	18:38	

5 Use the information in the table to work out the times in the different cities.

a It is 07:35 in London.

What time is it in New York?

City	Time difference	City
London	5 h ahead	New York
London	9 h behind	Sydney
Sydney	14 h ahead	New York

b It is 16:42 in London.

What time is it in Sydney?

c It is 11:58 in Sydney.

What time is it in New York?

d It is 21:17 in Sydney.

What time is it in London?

e It is 09:24 in New York.

What time is it in London?

Now look at and think about each of the *I can* statements.

Date: _____

Geometry and Measure

Name: _____

 Use the triangular dot paper to sketch these three triangles.

   a   scalene triangle – Label the triangle **A**.

   b   isosceles triangle – Label the triangle **B**.

   c   equilateral triangle – Label the triangle **C**.

 Name each of the triangles.

Three equal sides and three equal angles of 60°.	_____ triangle
No equal sides and no equal angles.	_____ triangle
Two equal sides and two equal angles.	_____ triangle

Geometry and Measure

 Draw the lines of symmetry on each pattern.

a

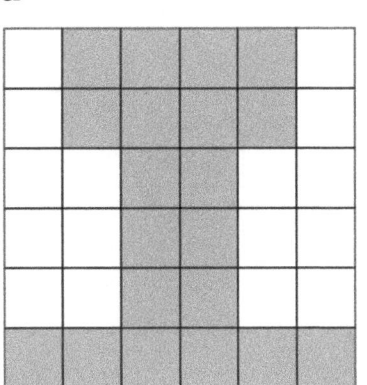

b

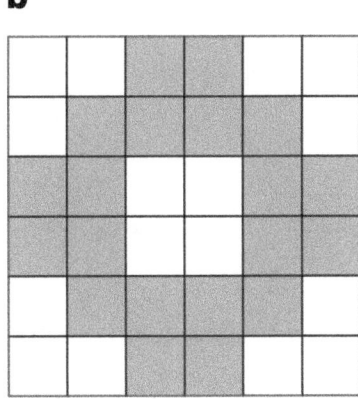

c

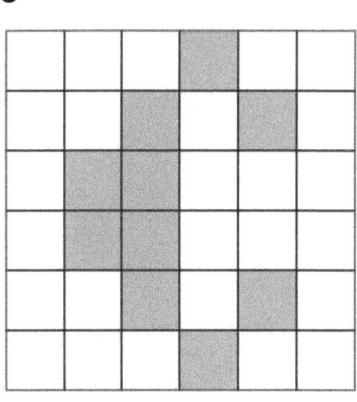

 Shade each grid to make a pattern with one or two lines of symmetry.

a

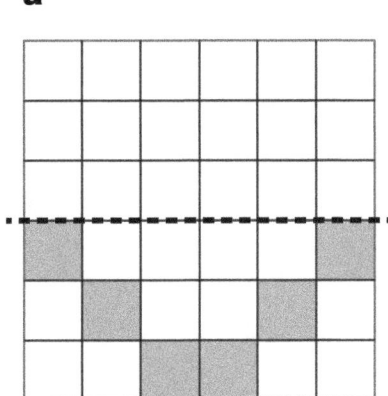

b

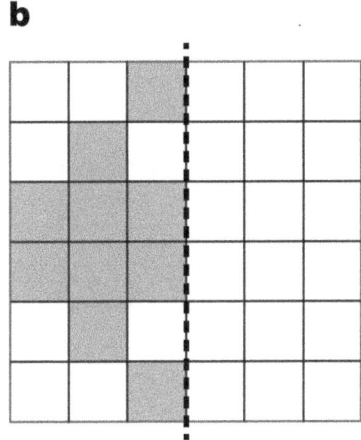

c

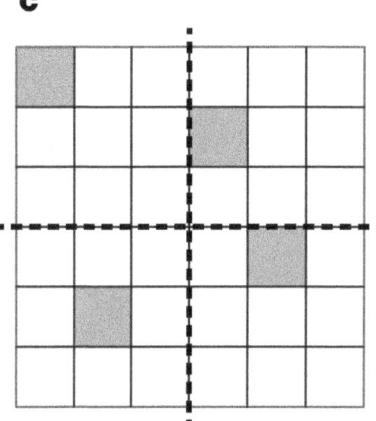

5 Create your own symmetrical patterns.

a

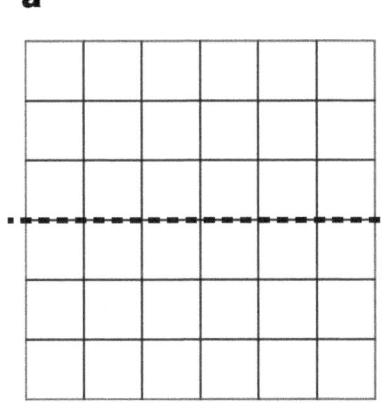

b

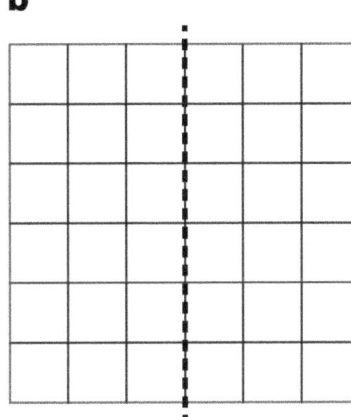

c

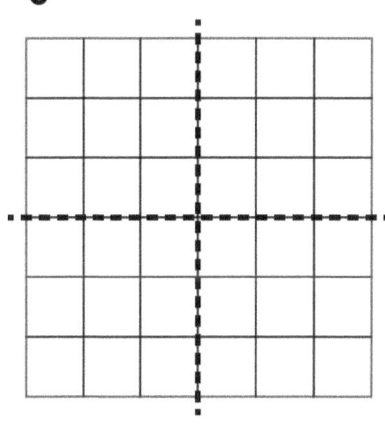

Now look at and think about each of the *I can* statements.

Date: _____

Name: _____

Geometry and Measure

1 Circle the polyhedra.

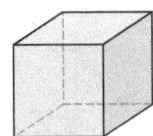 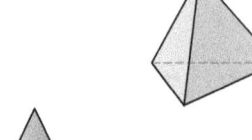  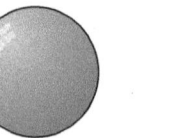 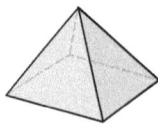

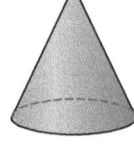 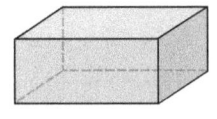 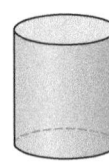

2 Use the triangular dot paper to sketch the two shapes.

  a cube – Label the cube **A**.

  b cuboid – Label the cuboid **B**.

3 Write the properties of these shapes.

  a Tetrahedron _____

  _____

  b Triangular prism _____

  _____

Geometry and Measure

4 Circle the shapes that are nets of an **open** cube.

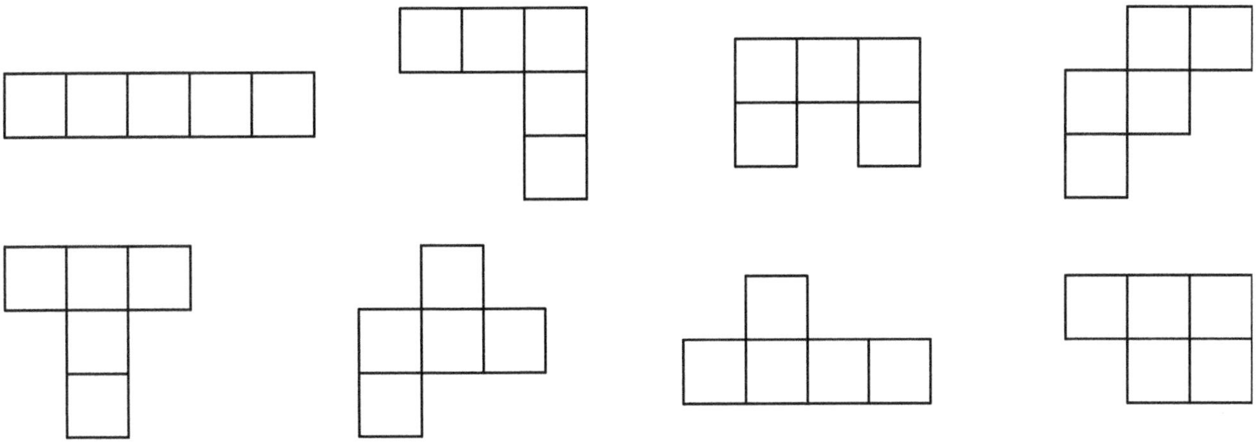

5 Circle the shapes that are nets of a **closed** cube.

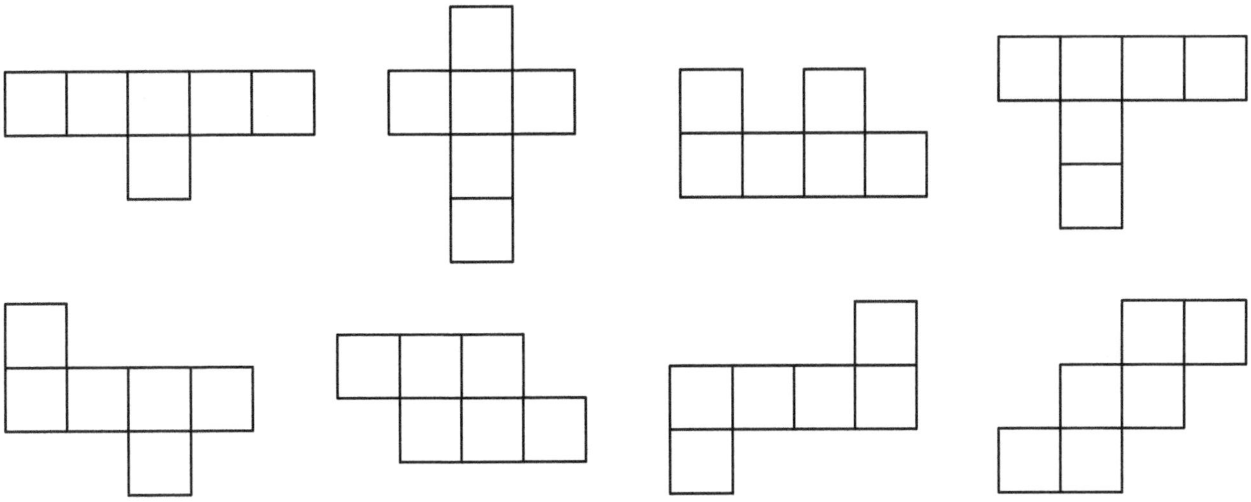

6 Sketch the net of a **closed** cube that has not been included in **5**.

Now look at and think about each of the *I can* statements.

Date: _____

Name: _____

 Identify and label the angles **acute**, **right**, **obtuse**, **straight** or **reflex**.

a

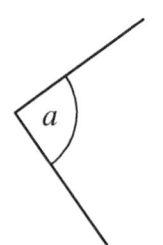

b

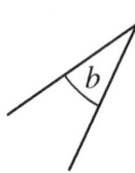

c

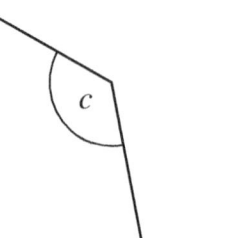

d

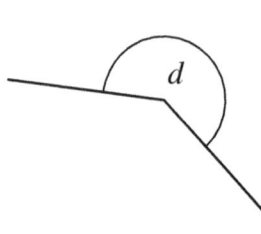

e

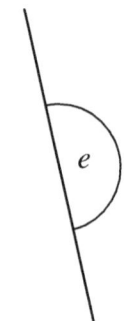

f

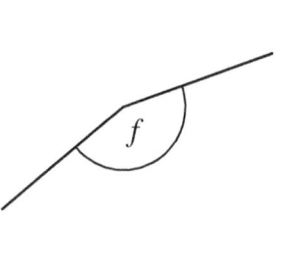

 Each angle has a letter. Place the angles in ascending and descending order by writing the letters in the boxes below.

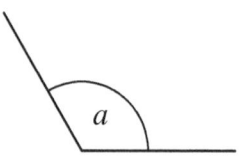

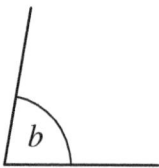

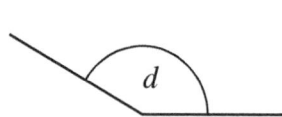

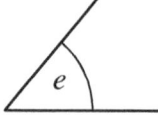

a Ascending order: ☐ < ☐ < ☐ < ☐ < ☐

b Descending order: ☐ > ☐ > ☐ > ☐ > ☐

Geometry and Measure

3 Write the unknown angles in the boxes. Remember to include the degrees symbol °.

a
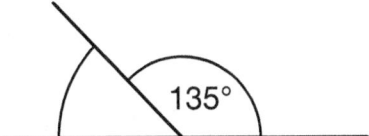
135°

b
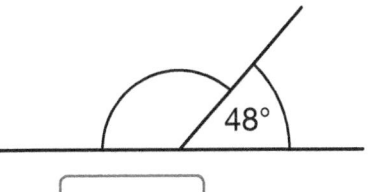
48°

c
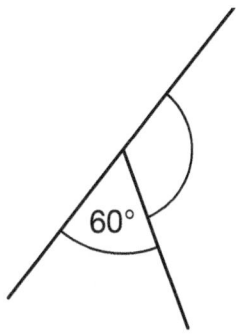
60°

d
112°

4 Write the unknown angles in the boxes. Remember to include the degrees symbol °.

a
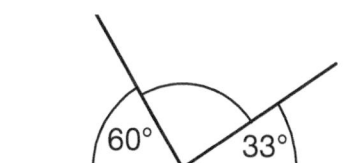
60° 33°

b
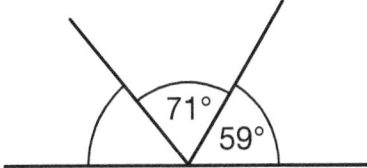
71° 59°

c
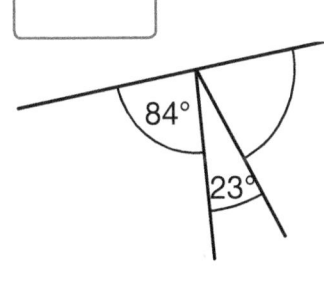
84° 23°

d
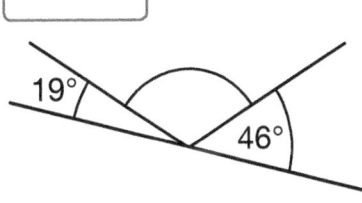
19° 46°

Now look at and think about each of the *I can* statements.

Date: _____

Name: _____

Geometry and Measure

You will need
• ruler

1 Calculate the missing dimension of each shape, given its perimeter and one dimension.

a P = 20 m

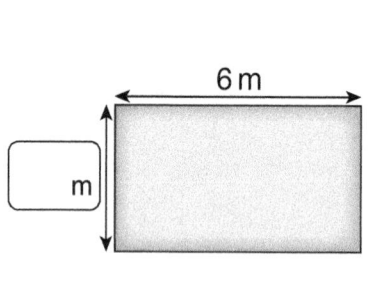

6 m

__ m

b P = 28 m

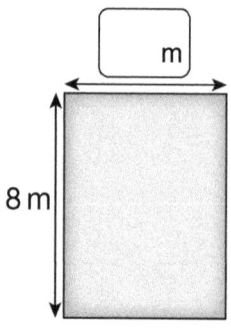

__ m

8 m

c P = 44 m

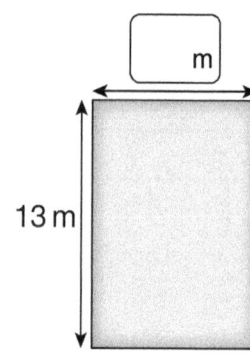

__ m

13 m

2 Calculate the perimeter of each shape.

a P = [____] m

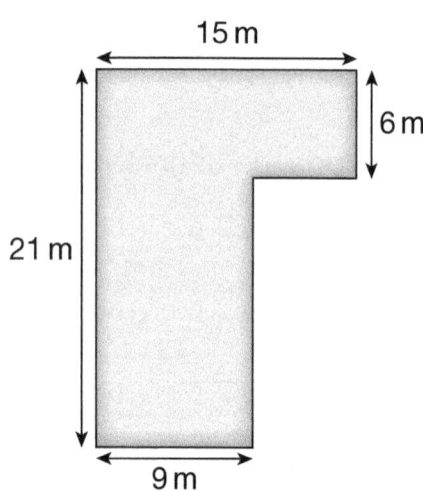

15 m

6 m

21 m

9 m

b P = [____] cm

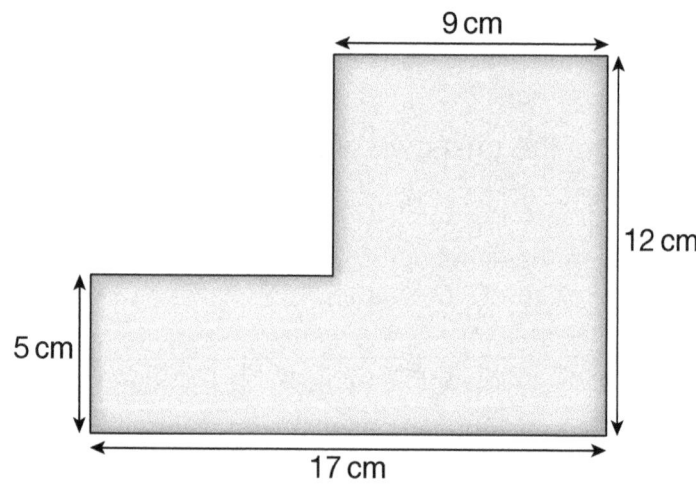

9 cm

12 cm

5 cm

17 cm

3 Calculate the perimeter and area of each shape.

a P = [____] cm

A = [____] cm²

9 cm

4 cm

b P = [____] cm

A = [____] cm²

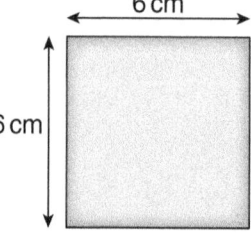

6 cm

6 cm

c What do you notice about the perimeters and areas of these shapes?

_____

4 You are given the area and one dimension of each rectangle.
Fill in the missing dimension.

a $A = 70\,m^2$

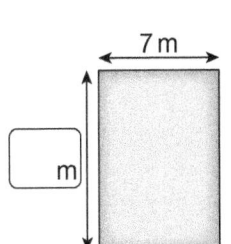

7 m

m

b $A = 72\,m^2$

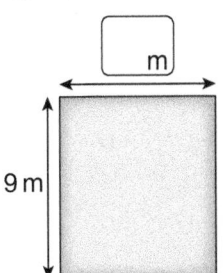

m

9 m

c $A = 84\,m^2$

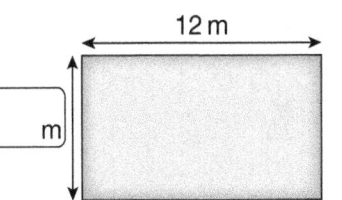

12 m

m

5 Calculate the area of each shape using your preferred method.
Show your working out.

a $A = \boxed{}\,m^2$

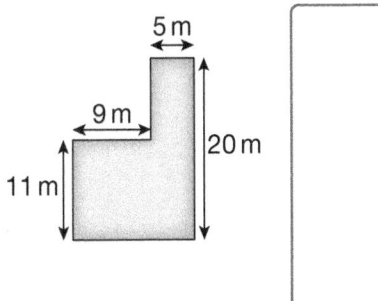

5 m

9 m

20 m

11 m

b $A = \boxed{}\,m^2$

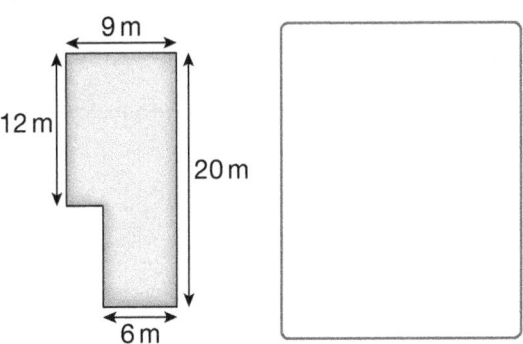

9 m

12 m

20 m

6 m

6 Draw and label a compound shape on the grid so that it has an area of $56\,cm^2$ and one dimension of 6 cm.

Now look at and think about each of the *I can* statements.

Date: _____

Name:

_____

1 Plot each pair of points on the coordinates grid, then circle the point that is closer to the *x*-axis.

a A (9, 4) B (3, 7)

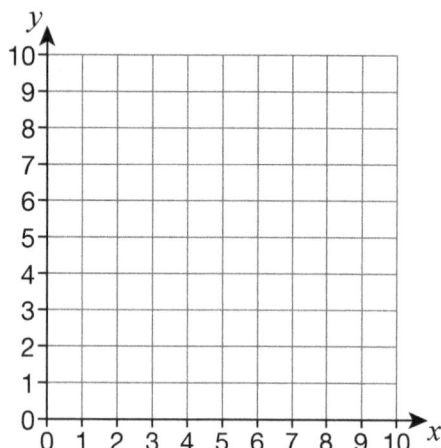

b C (1, 2) D (6, 5)

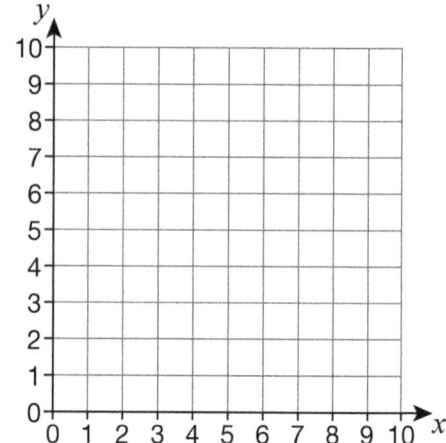

c E (7, 2) F (5, 3)

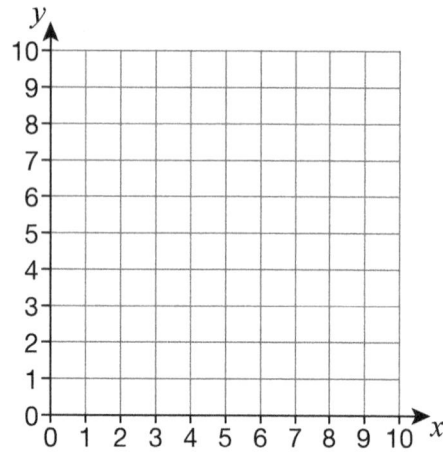

d G (2, 5) H (5, 2)

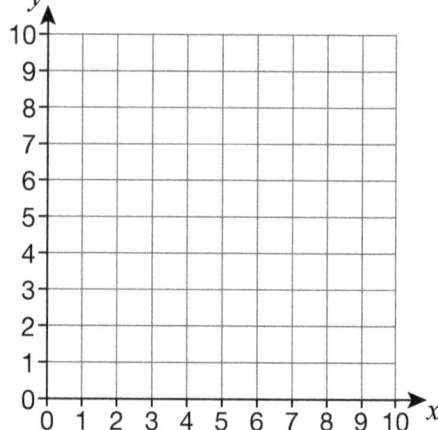

2 Plot each point.

a Point B (6, 3)

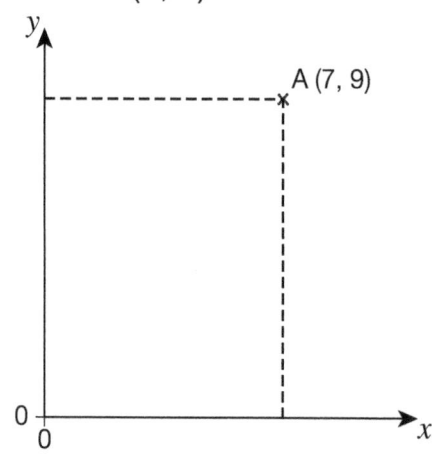

A (7, 9)

b Point D (8, 6)

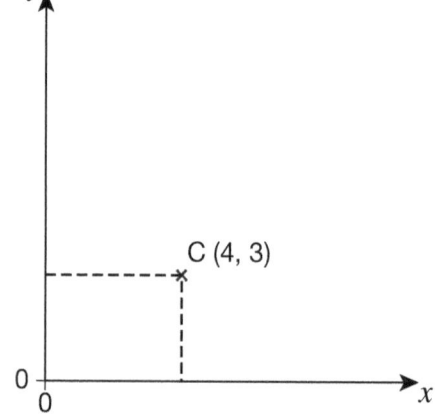

C (4, 3)

3 Draw each shape on the grid. Write the coordinates of each vertex of the shape.

a Draw a square. Start at point (2, 1).

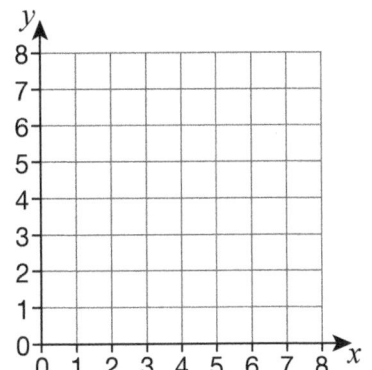

Coordinates of square: (2, 1) (—, —)
(—, —) (—, —)

b Draw a right-angled triangle. Start at point (1, 7).

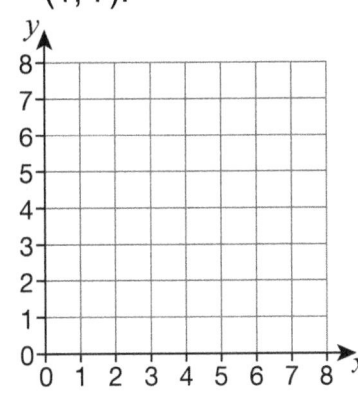

Coordinates of triangle:
(1, 7)
(—, —) (—, —)

c Draw a rectangle. Start at point (6, 3).

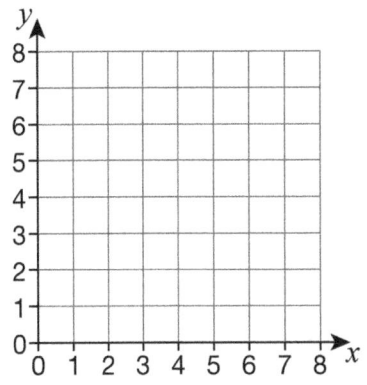

Coordinates of rectangle:
(6, 3)
(—, —) (—, —) (—, —)

 4 Draw each shape and identify the coordinates of the missing vertex.

a Plot these points:
A (1, 7) B (7,4). ABC is an isosceles triangle.

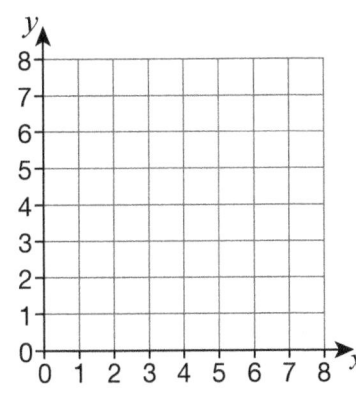

What are the coordinates of C?
(—, —)

b Plot these points:
A (2, 2) B (2, 7) C (7, 7). ABCD is a square.

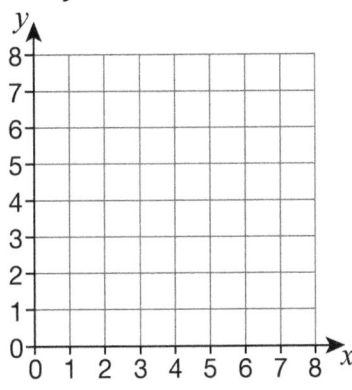

What are the coordinates of D?
(—, —)

c Plot these points:
A (1, 2) B (1, 6) C (6, 6). ABCD is a rectangle.

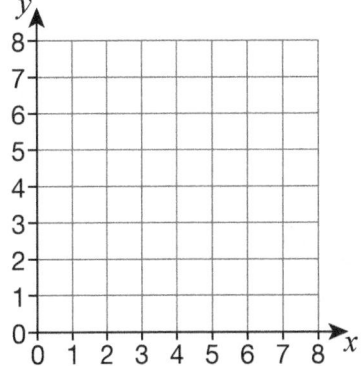

What are the coordinates of D?
(—, —)

Now look at and think about each of the
I can statements.

Date: _____

Geometry and Measure

Name: _____

1 Complete each sentence.

a Shape A has been translated ___ squares _____.

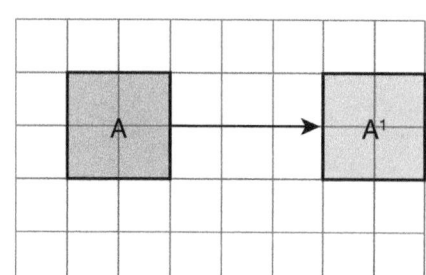

b Shape B has been translated ___ squares _____.

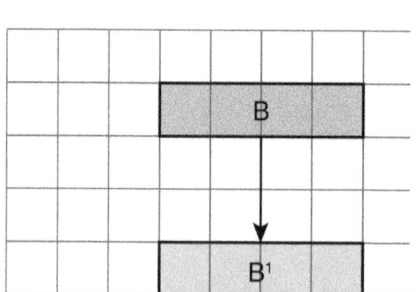

2 Describe the translation of each shape from A to A¹.

a _____

_____

_____

_____

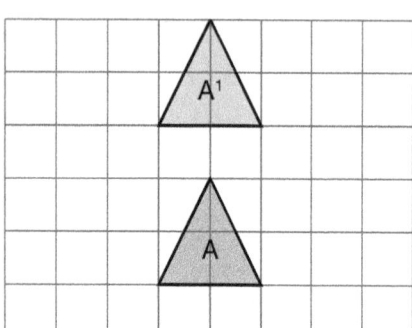

b _____

_____

_____

_____

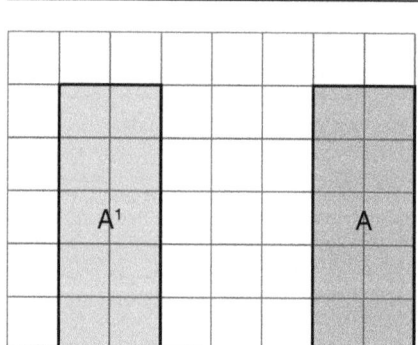

3 Translate each shape.

a Right 4 squares, up 3 squares

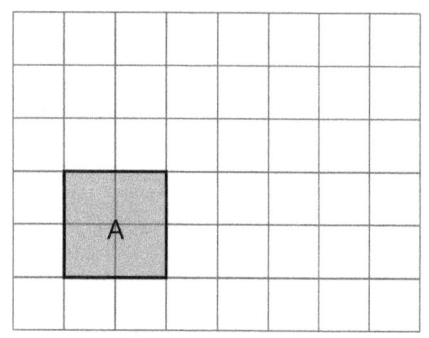

b Down 4 squares, left 5 squares

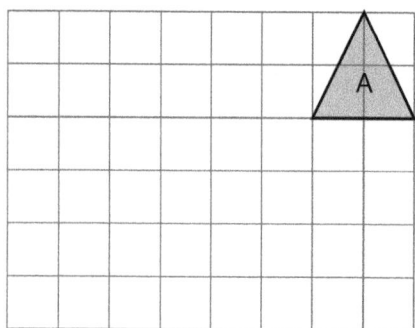

4 Reflect each shape in the mirror line.

a
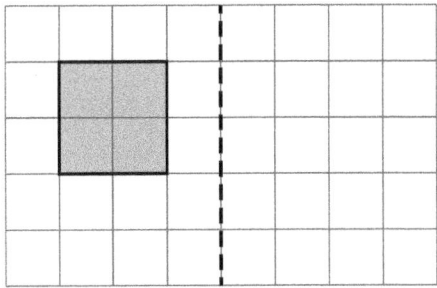

b
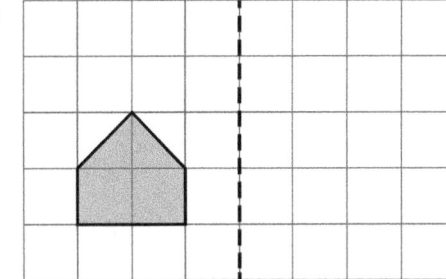

c
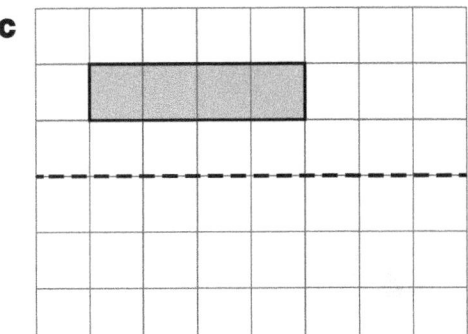

d
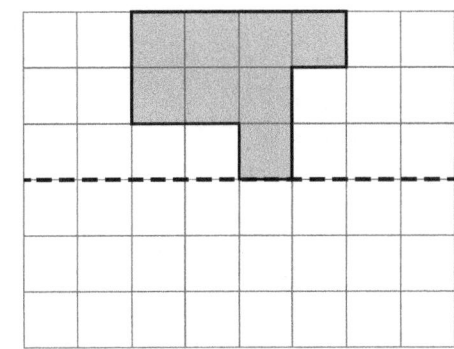

5 Reflect each shape in the mirror line.

a
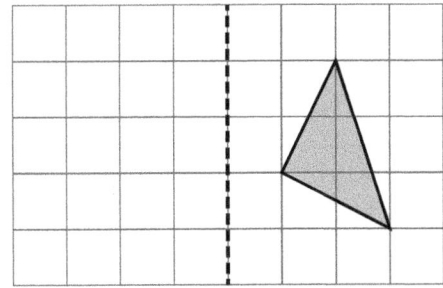

b
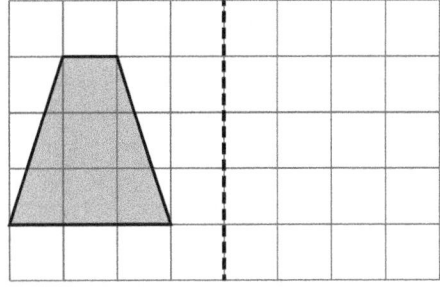

c
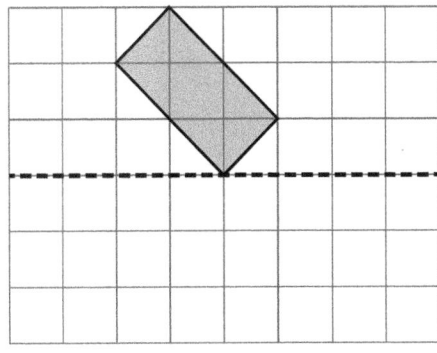

d
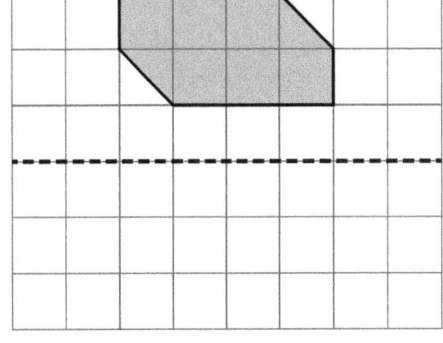

Now look at and think about each of the *I can* statements.

Date: _____

Statistics and Probability

Name: _____

1 Present the information in the Venn diagram in the Carroll diagram.

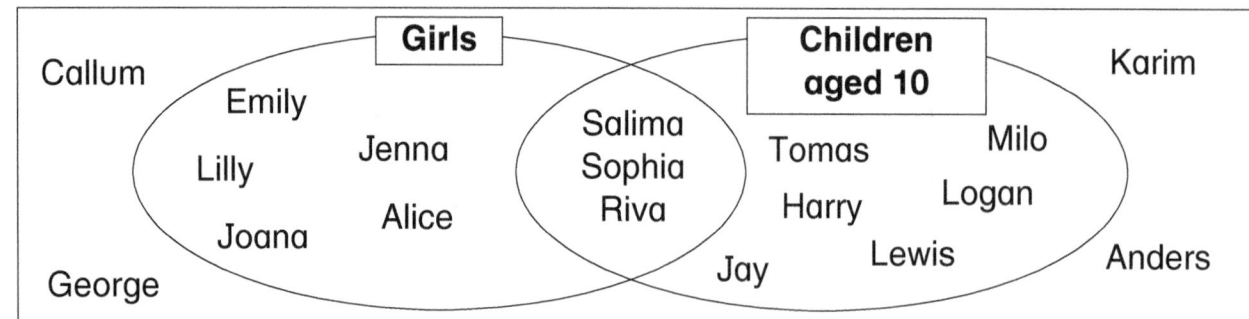

	Aged 10	Not aged 10
Girls		
Not girls		

2 Write three statements about the data in **1**.

i. _____

ii. _____

iii. _____

3 Write three statements about the data presented in the bar chart.

i. _____

_____

_____

ii. _____

_____

_____

iii. _____

_____

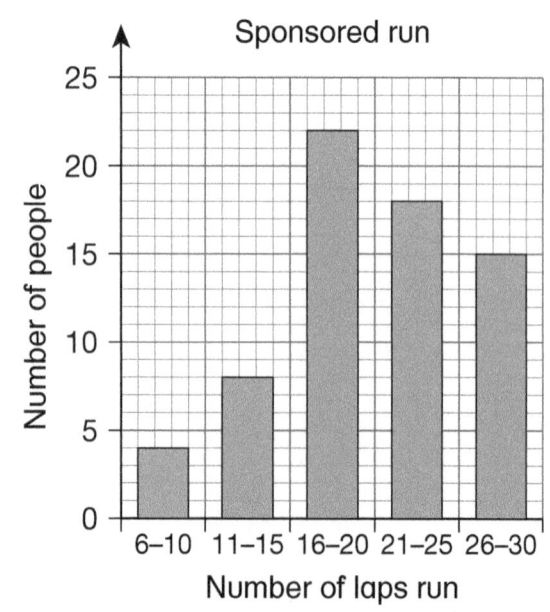

4 100 shoppers were asked the question, 'How did you get to the shopping centre today?' The data was recorded in a frequency table. Represent the data in a waffle diagram. The waffle diagram will need a title and a colour-coded key.

Transport	Frequency	Percentage
car	46	46%
bus	31	31%
walk	8	8%
bike	12	12%
other	3	3%

Key:

Statistics and Probability

5 What conclusions can you draw from the data in **4**?

_____

_____

_____

_____

6 Find the mode and median for each set of data.

a Spelling test scores: 10, 4, 5, 10, 6, 9, 10, 10, 5, 8, 10

Mode: ☐   Median: ☐

b Maths test scores: 18, 14, 16, 14, 15, 17, 15, 18, 20, 15, 19

Mode: ☐   Median: ☐

Now look at and think about each of the *I can* statements. ☐

Date: _____

Name: _____

You will need
- ruler
- coloured pencils

1 Class 5 took the temperature in their classroom at various times throughout the day and recorded the information in a table. Complete the line graph to display the data. Remember to label both axes.

Classroom temperature

Time	Temp °C
08:00	6
09:00	8
10:00	11
11:00	15
12:00	17
13:00	18
14:00	17
15:00	14

Classroom temperature

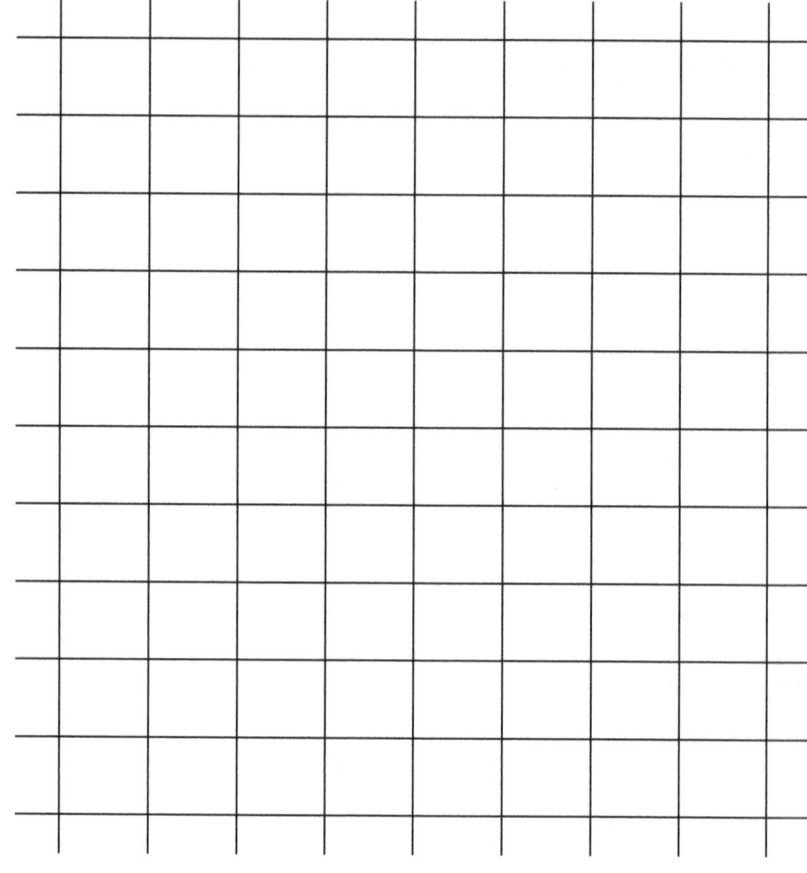

2 Write three statements about the data in **1**.

i. _____

ii. _____

iii. _____

3 William kept a record of the number of goals his favourite football team scored in a season. These are the results. Draw a tally chart or frequency table for the data.

0	2	1	3	2
1	4	0	0	1
2	1	4	3	2
0	2	1	1	2

4 Look at your completed tally chart or frequency table in **3**.
Draw a dot plot to display this data. Remember to label the axis.

5 Write three statements about the data in **4**.

i. _____

ii. _____

iii. _____

6 Look at the spinner. Choose from 'impossible', 'unlikely', 'even chance', 'likely' or 'certain' to describe each statement.

a   The chance of spinning 1: _____

b   The chance of spinning 2: _____

c   The chance of spinning an even number: _____

7 Colour each spinner using the colours blue, red and yellow only to give the chances described.

a   There is an even chance of spinning blue, but it is also possible to spin red and yellow.

b   The chance of spinning red is likely. The chance of spinning blue is unlikely. It is impossible to spin yellow.

Now look at and think about each of the *I can* statements.

Date: _____

The Thinking and Working Mathematically Star

Think about each of these *I can* statements and record how confident you feel about **Thinking and Working Mathematically**.

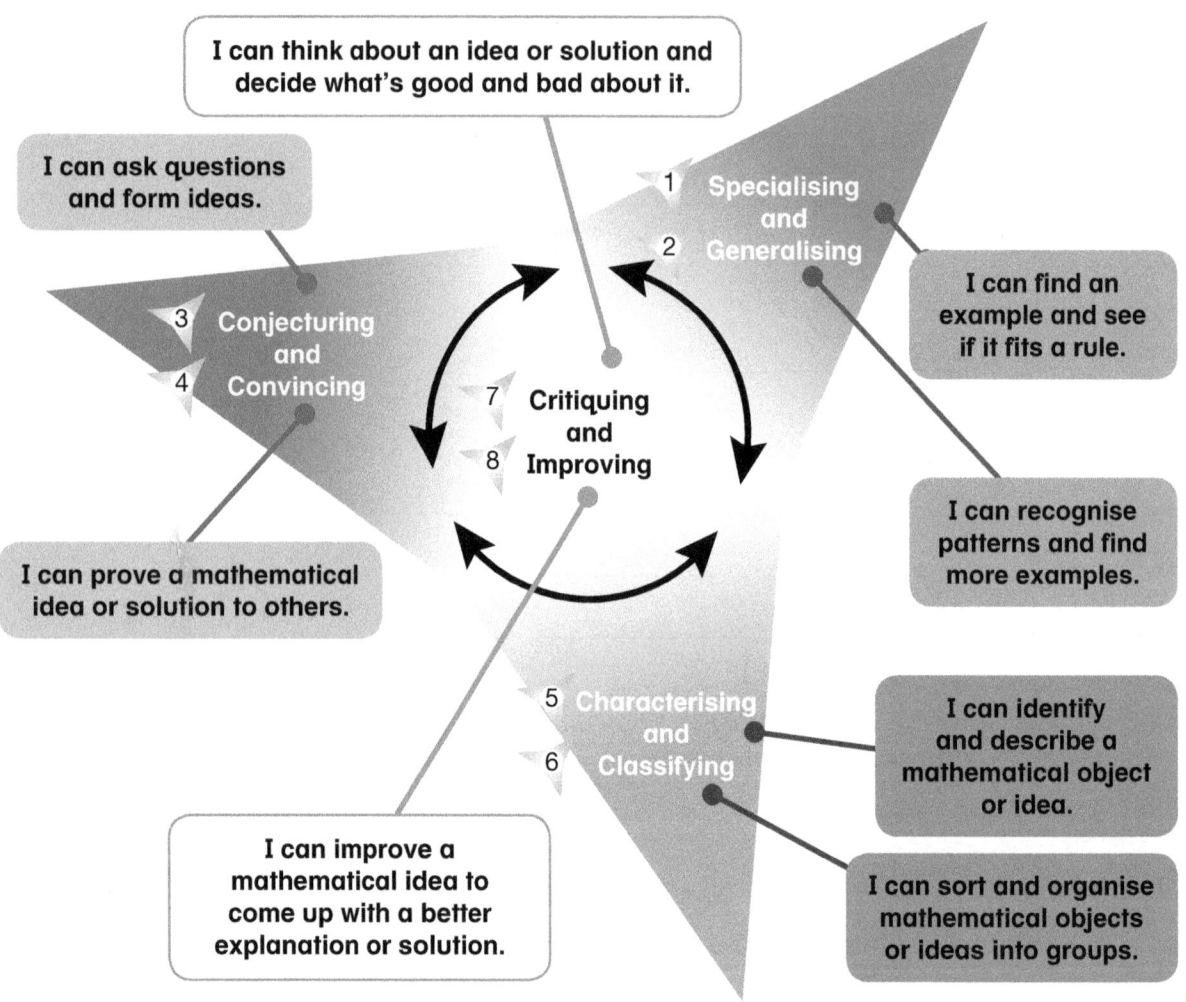

How confident do you feel Thinking and Working Mathematically?

Term ❶

Date: _____

☺ 😐 ☹

Date: _____

☺ 😐 ☹

Term ❷

Date: _____

☺ 😐 ☹

Date: _____

☺ 😐 ☹

Term ❸

Date: _____

☺ 😐 ☹

Date: _____

☺ 😐 ☹

The Thinking and Working Mathematically star, © Cambridge International, 2018